Mechanisms and Machine Science

Volume 121

This book series establishes a well-defined forum for monographs, edited Books, and proceedings on mechanical engineering with particular emphasis on MMS (Mechanism and Machine Science). The final goal is the publication of research that shows the development of mechanical engineering and particularly MMS in all technical aspects, even in very recent assessments. Published works share an approach by which technical details and formulation are discussed, and discuss modern formalisms with the aim to circulate research and technical achievements for use in professional, research, academic, and teaching activities.

This technical approach is an essential characteristic of the series. By discussing technical details and formulations in terms of modern formalisms, the possibility is created not only to show technical developments but also to explain achievements for technical teaching and research activity today and for the future.

The book series is intended to collect technical views on developments of the broad field of MMS in a unique frame that can be seen in its totality as an Encyclopaedia of MMS but with the additional purpose of archiving and teaching MMS achievements. Therefore, the book series will be of use not only for researchers and teachers in Mechanical Engineering but also for professionals and students for their formation and future work.

The series is promoted under the auspices of International Federation for the Promotion of Mechanism and Machine Science (IFToMM).

Prospective authors and editors can contact Mr. Pierpaolo Riva (publishing editor, Springer) at: pierpaolo.riva@springer.com

Indexed by SCOPUS and Google Scholar.

Med Amine Laribi · Giuseppe Carbone ·
Zhiyu Jiang
Editors

Advances in Automation, Mechanical and Design Engineering

SAMDE 2021

Springer

Editors
Med Amine Laribi
SP2MI—Site du Futuroscope
University of Poitiers
Poitiers, France

Giuseppe Carbone
DIMEG
University of Calabria
Rende, Italy

Zhiyu Jiang
iOPEN
Northwestern Polytechnical University
Xi'an, Shaanxi, China

ISSN 2211-0984 ISSN 2211-0992 (electronic)
Mechanisms and Machine Science
ISBN 978-3-031-09908-3 ISBN 978-3-031-09909-0 (eBook)
https://doi.org/10.1007/978-3-031-09909-0

This Springer imprint is published by the registered company Springer Nature Switzerland AG
The registered company address is: Gewerbestrasse 11, 6330 Cham, Switzerland

Contents

Chapter 1
Simulation and Research of the Final Phase of the Movement of the Descent Vehicle Landing Stage for the Conditions of Mars and Earth

Vsevolod V. Koryanov

Abstract The landing phase of a descent vehicle is a very crucial part of a mission. The most trustable landing technologies and systems developed are necessary in this stage to ensure the achievement of the main aims of the mission in the most accurate way possible. The environmental conditions of the landing impose restrictions on the possibility of conducting experimental studies, since it is rather difficult to simulate such conditions. Therefore, it is necessary to apply methods of numerical digital modeling. Conventionally, the landing of the landing vehicle can be divided into several stages-from the entry into the planet's atmosphere to the moment it approaches the surface, the moment of contact with the surface, and interaction with the surface. Different conditions of descent and environment at different stages impose peculiarities on the modeling process, and as a result, it is necessary to apply different modeling methods and different software. This article examines the final stage of the landing vehicle landing-the moment of contact with the surface and interaction with the planet's surface. An assessment is made of the influence of various conditions of approach to the surface on the magnitude of the arising overloads for the landing vehicle. The landing of the descent vehicle was simulated under various initial conditions and the values of the arising overloads on the descent vehicles were estimated.

Keywords Descent vehicle · Final phase · Overload · Payload · Numerical simulation · Different environmental conditions

V. V. Koryanov (✉)
Dynamics and Flight Control of Spacecraft and Rockets Department, Bauman Moscow State Technical University, Moscow, Russian Federation
e-mail: vkoryanov@bmstu.ru

M. A. Laribi et al. (eds.), *Advances in Automation, Mechanical and Design Engineering*, Mechanisms and Machine Science 121,
https://doi.org/10.1007/978-3-031-09909-0_1

1.1 Introduction

Currently, more and more countries and space agencies are planning their space missions to fly to other planets [1–3]. The complexity of such missions is created due to the presence of a large number of poorly studied environmental factors, the planet's surface and soil properties. As a consequence, experimental testing of such planetary environmental conditions in Earth conditions is very difficult, since it is difficult to simulate conditions for example on Mars. Among the missions that are currently in the active development stage, one can single out ExoMars [4–6]. The shape of the descent vehicle, its mass-geometric characteristics are given in the articles mentioned above. For comprehensive testing of all systems of such a spacecraft, digital modeling methods are used with the use of modern modeling tools. For example, we can talk about the use of a rather interesting software package for modeling the kinematics of mechanisms—the Universal Mechanism [7]. One of the important stages of a successful space mission is the landing of the descent vehicle on the planet's surface These issues can be distinguished in articles [8]. In addition, it is ineluctable to note other significant results obtained through numerical simulations by scientists regarding this field of scientific research [9–12].

Landing vehicles require special devices capable of adapting these overloads and forces to ensure a successful landing process. In order to better understand the changes in the interaction between the vehicle and the landing surface, it is necessary to conduct additional simulations varying the type of soil and the approach conditions, such as the inclination of the surface and the deviation of the vehicle axis in the moment of collision. From the entire landing process, it is necessary to separately distinguish the stage of contact with the surface—the most final stage of the movement when descending in the atmosphere. At this stage, it is very important to focus on the study of the resulting g-forces depending on the conditions of the approach to the surface. This research focuses on modeling and studying the movement of the descent vehicle at the final stage—at the moment of contact with the surface.

The main idea of modeling was formulated as follows—to simulate the movement of the descent vehicle in the soil of the planet—the surface as a spatial motion, but take into account that around the descent vehicle it is not the atmosphere but the soil with its own properties. This can be found in the works of the author [8, 13]. This article is part of a series of articles by the author dedicated to the landing process [14] and the study of the final stage of the movement [15, 16].

In this article, we focused on the following landing options: for the conditions of Mars, the conditions were set to an axis deviation (10°–20°), a landing site deviation (0°–25°) and two types of soil, hard and soft soil (see Fig. 1.1).

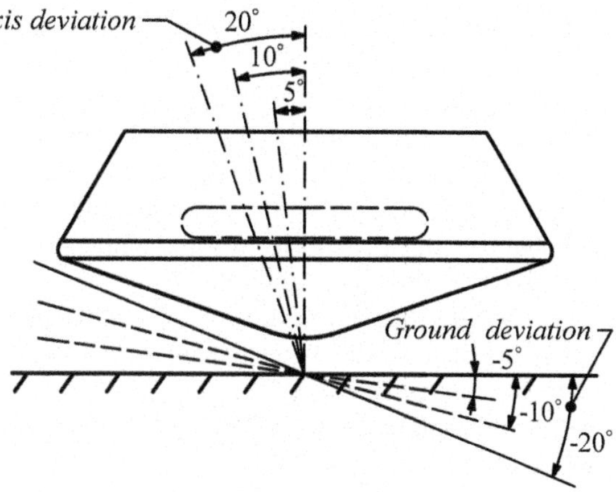

Fig. 1.1 Scheme of the landing vehicle with payload and initial conditions of modelling process

1.2 Vehicle Model

The landing vehicle is modelled as a conic body (Fig. 1.1), which produces a certain lift and drag during its descent motion. The shape and characteristics of the body are similar to the ones from the ExoMars mission vehicle. However, its dimensions and weight are such of a small satellite.

In Universal Mechanism, the model considered is a system of two completely rigid bodies: the landing device itself and the payload inside of it. The payload is modelled as a toroid (Fig. 1.1), which moves inside the landing vehicle following the classic mechanics theory, so the position of the centre of mass of the whole system changes according to the position of both bodies. The soil model was verified experimentally with analogs of natural soil rocks. The unevenness of the ground was taken into account due to the inclination of the landing site, in our case from −20 to +20° The modeling method showed good agreement with the experimental data carried out for similar descent vehicles and using the proposed digital modeling technique.

The model used in MATLAB, however, does not consider the existence of a separated rigid body for modelling the payload. Instead, it represents the payload as a rigidly attached part of the landing vehicle since its contribution to the motion of the main body is so little that it can be dismissed.

1.3 Landing Calculations Using Universal Mechanism and MATLAB Software

Universal Mechanism simulates kinematics and dynamics of planar and spatial mechanical systems, representing them as systems of rigid bodies connected by various kinematical pairs [7].

Simulations were conducted as if the landing took place on Earth and Mars (gravitational constant of 9.81 m/s^2 and 3.73 m/s^2 respectively) for two different types of soil: boulder and sand, which this paper will refer to as "hard soil" and "soft soil". Moreover, the symmetrical axis of the landing vehicle and the landing surface inclination varied for every simulation (Fig. 1.1), both in steps of 5°. The axis of the vehicle was deviated from the local vertical axis from 0° to 20°, while the surface inclined from the local horizontal axis a range from −20° to 20°. All in all, the study managed a total of 180 simulations.

Meanwhile, the landing area of the vehicle was calculated using MATLAB, since it analyses data iteratively and designs processes with a programming language that expresses matrix and array mathematics directly. Besides, it has a wide range of professional and rigorously tested toolboxes to facilitate the optimization of the code [17]. Earth was modelled using the object referenceEllipsoid, which represents the World Geodetic System of 1984 (WGS 84) reference ellipsoid [18], and Mars was represented with the Martian reference ellipsoid available in MATLAB's database. Gravity for both planets depending on the position of the vehicle was calculated using the function gravityzonal, which is based on the gravitational potential of the planet excluding its potential due to rotation [19]. The terrestrial atmospheric characteristics were calculated using the function atmoscoesa, a representation of the 1976 COESA United States standard lower atmospheric values [20]. However, to implement the characteristics of the Martian atmosphere, it was necessary to interpolate the experimentally obtained values of the Viking 1 profile data provided by NASA [21].

Lift and drag variation according to the velocity and angle of attack of the vehicle were obtained by digitalizing the graphs of the lift and drag coefficients found in papers from the Lavochkin Association [5, 6], using the online software WebPlotDigitizer (Version 4.2) [22]. The fluid density (ρ), provided by the atmospheric properties previously discussed, and velocity (V) that appear in Eqs. 1.1 and 1.2 changed for each altitude as the vehicle approached the surface. The aerodynamical surface (S) was stated to remain constant.

$$L = \frac{1}{2}\rho V^2 S C_L \tag{1.1}$$

$$D = \frac{1}{2}\rho V^2 S C_D \tag{1.2}$$

The force of gravity was implemented in the code following the law of universal gravitation. The mass of the vehicle (m) is attracted to the mass of the planet (M) according to the Eq. 1.3, considering as positive the direction along the radius ($\vec{r}$) from the mass M to the mass m. G is the gravitational constant and has an approximated value of $6.674 \cdot 10^{-11} m^3 kg^{-1} s^{-2}$. Since the vehicle has an insignificant mass in comparison with the mass of the planet, the gravitational parameter used in the code considered the mass of the vehicle negligible.

$$F_g = -G\frac{Mm}{r^3}\vec{r} \tag{1.3}$$

The trajectory of the vehicle under the influence of the previously described forces was obtained solving the differential equation provided by Newton's second law of motion (see Eq. 1.4), considering the vehicle mass constant during the landing process.

$$\sum F = F_g + L + D = \frac{dH}{dt} = \frac{d(mV)}{dt} = m\frac{dV}{dt} \tag{1.4}$$

To solve the Eq. 1.4, it was required to establish the initial conditions of the movement of the vehicle: its initial position and its initial velocity expressed in the geographical system of coordinates. The initial velocity of the vehicle was set to be parallel to the Equator, with an initial pitch angle that could be modified for every simulation. After setting the initial parameters of the system, the differential equation was solved using the MATLAB function ode45, a function that iterates the code through a medium order method [23] until the final position is achieved (the vehicle intersects the surface of the planet).

1.4 Simulation Results and Discussion

Let us present the graphs and the main results for the case of modeling the final stage of the descent vehicle landing under the following conditions (Fig. 1.1): deviation of the longitudinal axis of the descent vehicle 10°, deviation of the landing site 20°, 15°, 10°, 5°, 0°, −5°, −10°, −15°, −20°, two types of soil of the planet's surface: soft and hard. Let's simulate for two environmental models: Earth and Mars. In addition, we will show the results separately for the lander body and the payload container.

1.4.1 Simulations on Earth

In Fig. 1.2, it can be noticed that the maximum value of the overload is found in the −10° surface inclination simulation, followed closely by the −5° surface inclination simulation. However, the results from the simulations with surface inclination angles of −20°, −15°, −10° and −5° should not be dismissed, since they present three (in the case of the −20° inclination) and two peaks (in the rest of inclinations mentioned). Therefore, neither of these simulations would represent a safe landing option for the vehicle. Paying attention to the time the overload takes to dissipate, it could be said that the best landing options would happen for surface inclination angles of 10°, 15° and 5°.

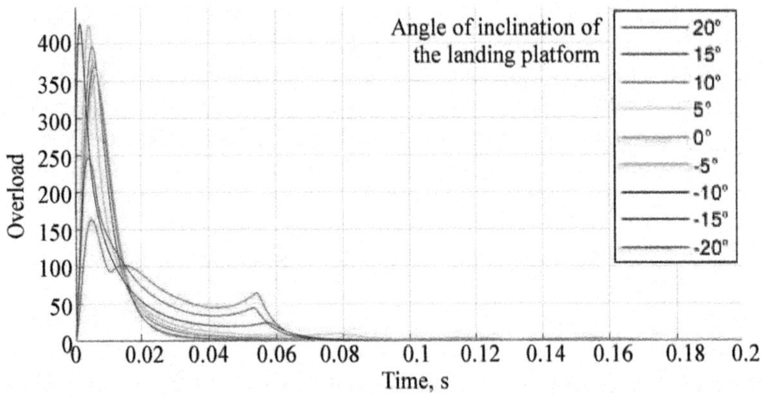

Fig. 1.2 Overload for different landing surface inclinations acting on the landing vehicle (10° body axis inclination) for soft soil on Earth

Nevertheless, Fig. 1.3 shows some changes in the values of the overloads compared to Fig. 1.2, which should be taken into consideration, since they represent the overload acting on the payload. The overload with the greatest amplitude happens for a surface inclination angle of 20°, followed by the results for angles 15° and 10°. The second and third peaks happen in the same cases as in Fig. 1.2, so they should be avoided in order to protect the payload from damage.

All in all, for soft soil landings on Earth with a body axis inclination of 10°, it could be concluded that the safest landings (for both the vehicle and the payload) would occur for surface inclination angles of 10° and 15°, while the most dangerous ones would occur for angles of −20°, −15°, −10°, −5° and 20°.

Therefore, for hard soil landings on Earth with a body axis inclination of 10°, it could be concluded that the safest landings (for both the vehicle and the payload)

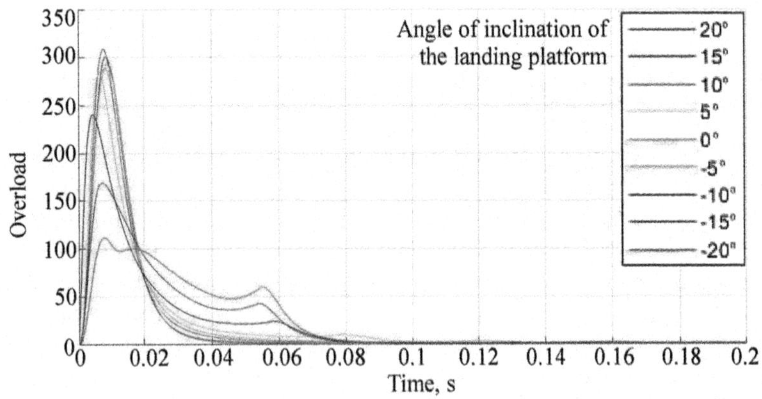

Fig. 1.3 Overload for different landing surface inclinations acting on the payload (10° body axis inclination) for soft soil on Earth

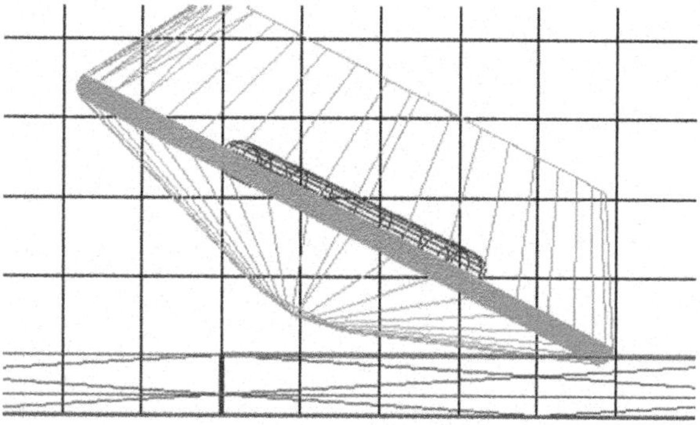

Fig. 1.4 Landing on a soft soil 0° inclined surface on Earth with a 10° vehicle axis deviation

would occur for a surface inclination angle of 15°, while the most dangerous ones would occur for angles of −20°, −15°, −10°, −5°, 0° and 20°. The visualization of process showed at Figs. 1.4, 1.5, 1.6 and 1.7.

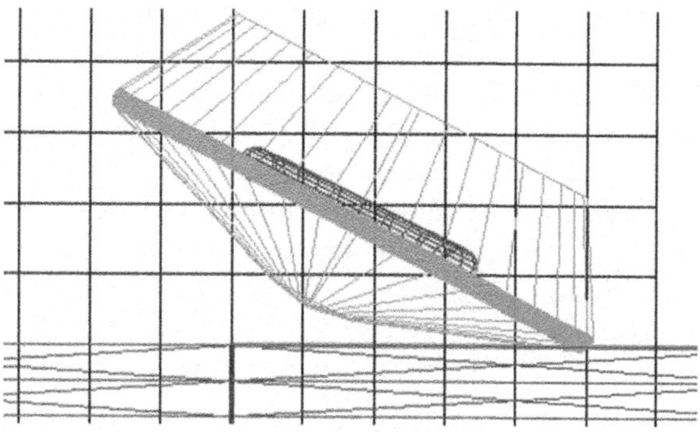

Fig. 1.5 Landing on a hard soil 0° inclined surface on Earth with a 10° vehicle axis deviation

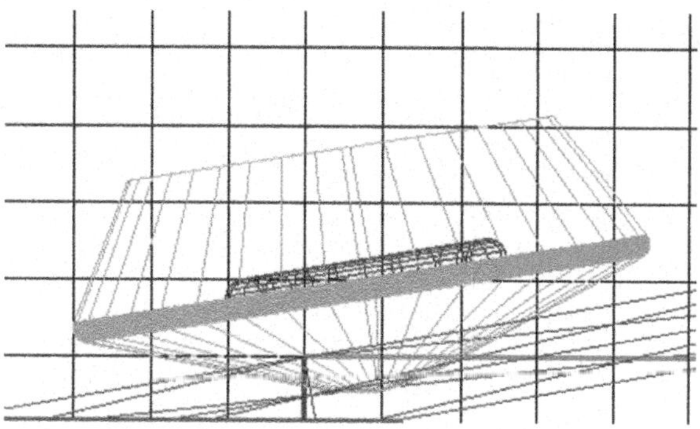

Fig. 1.6 Landing on a soft soil 10° inclined surface on Earth with a 10° vehicle axis deviation

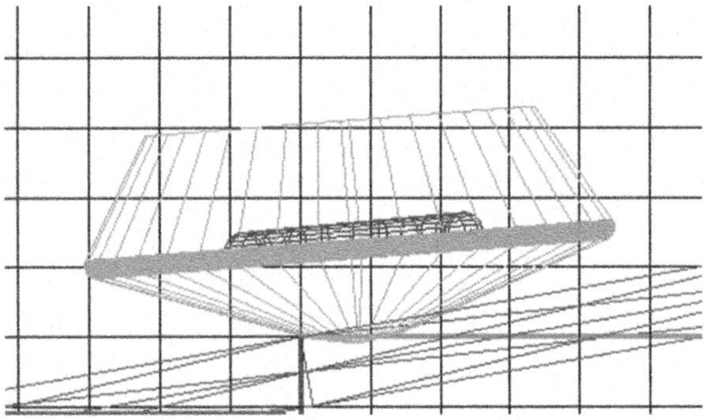

Fig. 1.7 Landing on a hard soil 10° inclined surface on Earth with a 10° vehicle axis deviation

1.4.2 Simulations on Mars

For landings on soft soil on Mars, the maximum value of the overload acting on the main body is achieved for the −5° surface inclination simulation (see Fig. 1.8), followed closely by the 20° surface inclination. However, it should be noticed that the −20° and −15° inclination curves slightly deviate from the common shape that the other inclinations have, showing the foundation of possible second peaks. Therefore, these simulations should be discarded as safe landing options. The simulations that take the littlest time to dissipate the overload are the 10° and 15° simulations.

The maximum value of the overload acting on the payload, however, is obtained for the 10° surface inclination simulation (see Fig. 1.9), followed by the one obtained for the 15° surface inclination simulation. In this case, the −20° and −15° inclination simulations also show the foundation of possible second peaks, and the fastest simulations that reach an overload value of zero are also the same.

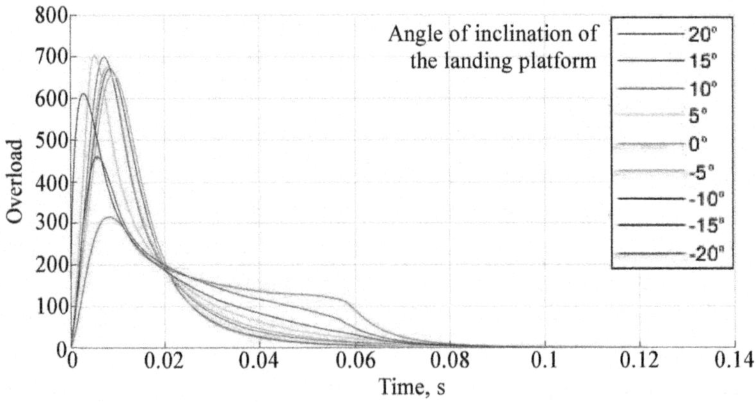

Fig. 1.8 Overload for different landing surface inclinations acting on the landing vehicle (10° body axis inclination) for soft soil on Mars

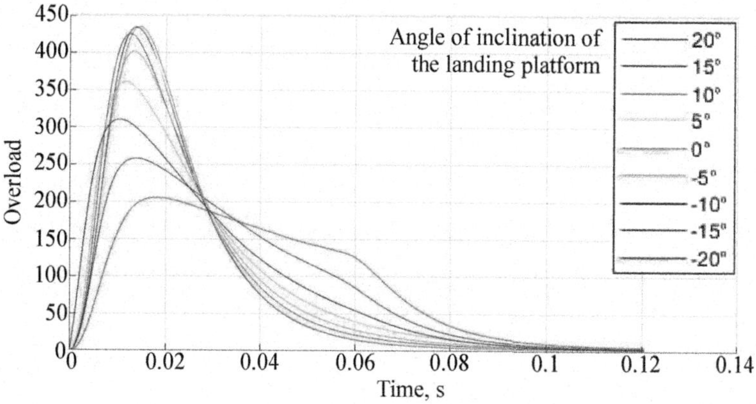

Fig. 1.9 Overload for different landing surface inclinations acting on the payload (10° body axis inclination) for soft soil on Mars

As a conclusion, for soft soil landings on Mars with a body axis inclination of 10°, the safest landing (for both the vehicle and the payload) would happen for the surface inclination angle of 15°. This is due to the fact that the payload maximum overload value is achieved for the 10° surface inclination simulation, as well as the littlest time the overload takes to dissipate. Therefore, it would be a more reliable solution to set the 15° surface inclination simulation as the safest simulation, since its overload value is smaller. On the contrary, the most dangerous landings would occur for angles of −20° and −15°.

In summary, the safest landing that could be chosen for hard soil on Mars with a body axis inclination of 10° would be the landing on the 10° inclined surface. By contrast, the more dangerous landings would happen for −20°, −15°, −10°, −5°, 0° and 20° surface inclinations. The visualization of process showed at Figs. 1.10 and 1.11.

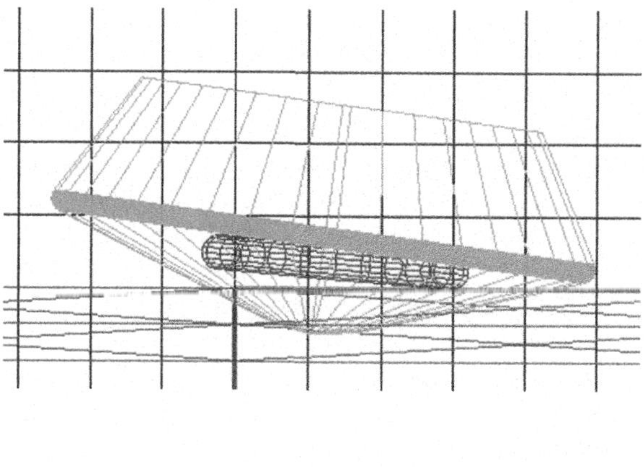

Fig. 1.10 Landing on a soft soil 0° inclined surface on Mars with a 10° vehicle axis deviation

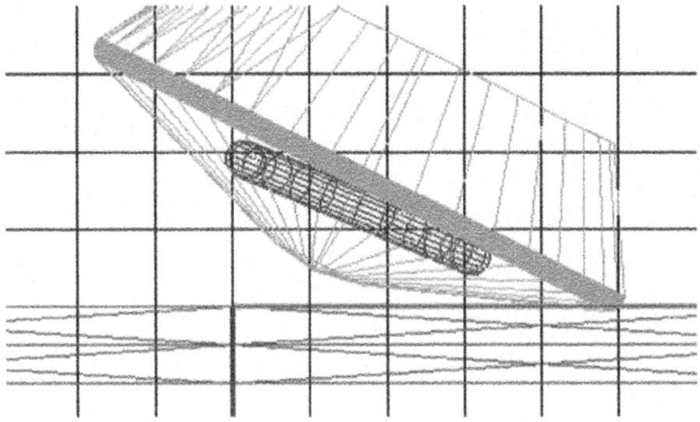

Fig. 1.11 Landing on a hard soil 0° inclined surface on Mars with a 10° vehicle axis deviation

1.5 Conclusion

As a result of the work performed, it can be concluded that the process of landing on the surface of a planet with deliberately poorly studied properties of the environment is very difficult for the successful implementation of a space mission. It is necessary to divide the landing process into stages of movement: entry into the atmosphere, movement in the atmosphere and the moment of approaching the surface—the moment of contact with the surface.

To analyze the dynamics of the descent vehicle at the stage of contact with the surface, it is possible to apply the calculation method described in this article. It is based on the assumption that the descent vehicle in the soil of the planet moves in the same spatial way as in the atmosphere of the planet, only there is soil around the descent vehicle.

This stage of movement of the descent vehicle strongly depends on the conditions of approach to the surface, the deviation of the longitudinal axis of the descent vehicle and the deviation of the landing site of the descent vehicle. The greatest values of the overloads, as shown by the calculations, will be with the greatest difference in the angles of deflection of the longitudinal axis of the descent vehicle and the landing site. In addition, the type of soil on which the landing takes place certainly has an impact, so the landing site for future missions must be carefully selected. However, it must be said here that it is necessary to choose which option is more preferable—a longer process of the influence of overloads but lower values of overloads for soft soil, or a faster process of influence of overloads but large values for hard soil. All these factors and the modeled landing options must be taken into account by the designers when developing new space missions.

As a result of the work performed, conclusions were drawn and a comparison of the arising overloads for the descent vehicle for landing in the conditions of Earth and Mars for the landing variant of the deviation of the longitudinal axis of the descent vehicle 10° and the deviation of the landing site from −20° to +20°. The most dangerous landing options were identified in terms of the magnitudes of the arising overloads.

References

1. ExoMars mission official web site. http://www.esa.int/Science_Exploration/Human_and_Robotic_Exploration/Exploration/ExoMars
2. InSight Mission—NASA's InSight Mars Lander. https://mars.nasa.gov/insight/
3. Martian Moons eXploration. http://mmx.isas.jaxa.jp/en/
4. Robotic Exploration of Mars—ESA. https://exploration.esa.int/web/mars/
5. Golomazov, M.M., Finchenko, V.S.: Aerodynamic design of descent vehicle in the martian atmosphere under "ExoMars" project. Institute of Automatics and Design, p. 42. Lavochkin Association. Moscow (2013)
6. Finchenko, V.S., Ivankov, A.A., Shmatov, S.I., Mordvinkin, A.S.: The preliminary results of the calculation and test investigations of the main aero and thermodynamic parameters "ExoMars" descent module, pp. 68–70. Lavochkin Association, Moscow (2013)
7. Universal Mechanism. Home Page. 2021. http://www.universalmechanism.com/en/pages/index.php?id=1
8. Koryanov, V.V., Kazakovtsev, V.P.: Dynamics of angular motion of landing vehicle in Martian atmosphere with allowance for small asymmetries. Int. J. Mech. Eng. Robot. Res. **7**(4), 385–391 (2018). https://doi.org/10.18178/ijmerr.7.4.385-391
9. Nerovny, N.A., Zimin, V.N.: Light radiation pressure upon a wrinkled membrane—Parametrization of an optically orthotropic model. J. Phys.: Conf. Ser. **991**(1) (2018), Article number 01206: 25th International Conference on Topical Problems of Continuum Mechanics, TPCM 2017. Tsakhkadzor, Armenia (2017). https://doi.org/10.1088/1742-6596/991/1/012062.
10. Zikun, Z., Zimin, V.N., Krylov, A.V., Churilin, S.A.: Investigation of dynamics of transformable space structure movement (Conference Paper). In: AIP Conference Proceedings, vol. 2103 (2019); 10th All-Russian Conference on Fundamental and Applied Problems of Modern Mechanics, FAPMM 2018. National Research Tomsk State University Tomsk, Russian Federation (2018). https://doi.org/10.1063/1.5099884
11. Zimin, V.N., Zikun, Z.E. Krylov, A.V., Churilin, S.A.: Mathematical modeling of the deployment of a large transformable space structure (Conference Paper) (Open Access). In: AIP Conference Proceedings, vol. 2171 (2019); 43rd Academic Space Conference: dedicated to the Memory of Academician S.P. Korolev and Other Outstanding Russian Scientists—Pioneers of Space Exploration ""Korolev Academic Space Conference", vol. 9. https://doi.org/10.1063/1.5133168
12. Zarubin, V.S., Zimin, V.N., Kuvyrkin, G.N.: Temperature distribution in the spherical shell of a gauge-alignment spacecraft. J. Appl. Mech. Tech. Phys. **58**, 1083–1090 (2017). https://doi.org/10.1134/S0021894417060141
13. Koryanov, V.V., Kazakovtsev, V.P., Griselin, H., Danhe, C.: Analysis of the dynamics of movement landing vehicle with an inflatable braking device on final stage of trajectory. IOP Conf. Ser. Mater. Sci. Eng. 3–8 (2019). https://doi.org/10.1088/1757-899X/630/1/012005
14. Koryanov, V.V., Richier, L., Chen, D.: Dynamics of the landing vehicle during the entry in the high atmosphere of mars using an inflatable braking device. In: 2020 8th International Conference on Control, Mechatronics and Automation, ICCMA 2020, pp. 131–137, 9301507 (2020). https://doi.org/10.1109/ICCMA51325.2020.9301507
15. Koryanov, V.V., Hermosilla Heras, M.: Research dynamic motion of a landing vehicle during the final landing stage. IOP Conf. Ser.: Mater. Sci. Eng. **904**(1). https://doi.org/10.1088/1757-899X/904/1/012014
16. Koryanov, V.V., Hermosilla, M.: Use of modern digital software to model the motion dynamics of spacecraft during landing. In: ITM Web of Conferences. vol. 35. EDP Sciences (2020). https://doi.org/10.1051/itmconf/20203504012
17. MathWorks.: Products. MATLAB. Overview (2019). https://www.mathworks.com/products/matlab.html
18. MathWorks.: MATLAB. Documentation. Reference ellipsoid (2012)
19. MathWorks.: MATLAB. Documentation. Gravityzonal (2009)

20. MathWorks.: MATLAB. Documentation. Atmoscoesa (2006)
21. Braeunig, R.A.: Martian Atmosphere. Rocket and Space Technology (2019). http://www.braeunig.us/space/atmmars.htm
22. Rohatgi, A.: WebPlotDigitizer. Version 4.2 (2019). https://automeris.io/WebPlotDigitizer
23. MathWorks.: MATLAB. Documentation. ode45 (2006)

Chapter 2
Heterogeneous Multi-robot Adversarial Patrolling Using Polymatrix Games

Alexander Langley, Vikas Dhiman, and Henrik Christensen

Abstract We consider the problem of patrolling an area with a heterogeneous multi-robot team. The goal for the patrolling robots is to detect intrusions while minimizing the probability of being detected. The heterogeneous teaming problem has been widely studied using coordination strategies such as minimax and strategic alliances. In most cases a centralized coordination is assumed. We present an approach based on a zero- sum polymatrix approach and derive its basic characteristics. In polymatrix games a patroller consider other team members' possible strategies when formulating their own strategy, while eliminating the requirement of considering the joint action space. We systematically evaluate our approach in randomly generated gridworld environments, and also demonstrate the performance of the approach in a large scale realistic environment. The results clearly demonstrate that the polymatrix approach has improved detection rates without an increase in complexity.

Keywords Multi-robot patrolling · Adversarial patrolling · Polymatrix games · Distributed patrolling

2.1 Introduction

Multi-robot deployment for security applications is becoming pervasive. We see multi-robot teams used for monitoring forest fires [1], for border control inspection [2], and for monitoring infrastructure [3] and bodies of water such as lakes [4]. The deployment of multiple autonomous vehicles can be formulated as a patrolling problem, where a multi-robot team of patrollers seeks to detect intruders in a large area [5, 6]. Patrolling, like the closely related Perimeter Defense [7] and Pursuit-Evasion [8, 9] problems, has applications in domains such as military operations [3], search and rescue [10], and checking in public transportation systems [11].

In this paper, we are primarily interested in a scenario in which a heterogeneous multi-robot team must provide reconnaissance of a large area in which intrusion

A. Langley (✉) · V. Dhiman · H. Christensen
University of California San Diego, San Diego, CA, USA
e-mail: alangley@ucsd.edu

M. A. Laribi et al. (eds.), *Advances in Automation, Mechanical and Design Engineering*, Mechanisms and Machine Science 121,
https://doi.org/10.1007/978-3-031-09909-0_2

has already occurred. The intruding robot's mission is to occupy and surveil the environment, but its exact type (ground or aerial) and position are not known to the patrolling team. The patrolling robots' goal in this situation is twofold: they seek to detect the intruder, but they also want to minimize the chance that they will be detected themselves.

To this end, we envision a team of ground and aerial robots that can patrol large-scale environments. In large-scale environments, the robots may not be able to maintain connectivity with each other at all times, which makes decentralization a necessity. However, most works in the literature model adversarial patrolling as a two-player game, which cannot be easily implemented in a decentralized fashion.

To address this limitation, we propose the use of polymatrix games [12] for adversarial patrolling. Polymatrix games allow us to model each patroller as a separate player who formulates their strategy given information about the other robots on their team. This formulation has the potential to be implemented as a decentralized algorithm on each robot. In this work, we present and evaluate a centralized version of the algorithm, leaving the decentralized version for future work. The main objective is to demonstrate the value of polymatrix games and for that the increased complexity of distributed coordination is not required.

On the other extreme, the polymatrix game is superior to the naïve solution of playing pairwise independent minimax games. Patrolling using polymatrix games accounts for the possible strategies of other patrollers, which is important because the exact composition of the patrolling team affects the intruder's best response. Without considering the possible strategies of their teammates, patrolling robots will make less informed predictions about how intruders will behave. Hence, our approach is a step toward decentralized agents who are also able to reason about the possible actions of their teammates.

Figure 2.1 summarizes our approach: Given an environment, we extract a graph representation. A team of autonomous robots comprised of ground and aerial robots then plays a polymatrix game with the intruder to decide where each agent will patrol and how often.

The main contribution of this work is to apply the polymatrix game formulation to the problem of multi-robot adversarial patrolling with a heterogeneous team of robots.

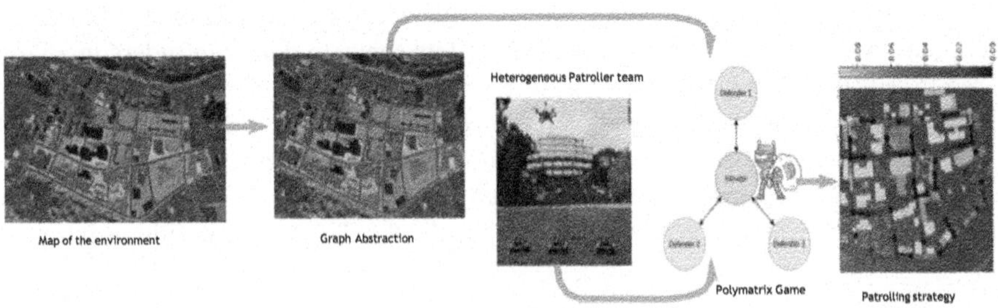

Fig. 2.1 Flowchart for generating adversarial patrolling strategies using polymatrix games

To the best of our knowledge, our work is the first to apply zero-sum polymatrix games to the adversarial patrolling problem.

The rest of this paper is organized as follows: In Sect. 2.3, we discuss our problem formulation. In Sect. 2.4, we introduce polymatrix games, discuss their application to multi-robot patrolling, and show how to compute patrolling strategies by finding their Nash equilibria. In Sect. 2.7, we evaluate our approach experimentally in randomly generated gridworld environments. In Sect. 2.8, we discuss the application of our approach to a realistic, large-scale environment.

2.2 Related Work

Patrolling is generally classified as either Regular or Adversarial patrolling [6]. Regular patrolling [5] aims to periodically visit important locations so that the duration between visits to locations is minimized. Regular patrolling, however, is typically deterministic. In the presence of an adversary which is able to observe the patrollers' behavior before deciding on a plan for intrusion, deterministic patrolling allows an intruder to invade a location when it knows that patrollers will be elsewhere. Adversarial patrolling addresses this problem by using stochastic strategies that resist the intruder's ability to learn and predict the patrollers' actions [6].

Adversarial patrolling is often formulated and solved in the literature as a two-player Bayesian Stackelberg game [6, 13–15]. In a Stackelberg game the leader (in the case of adversarial patrolling, the patroller team) chooses a strategy and the follower (here an intruder) observes and adapts to the leader's strategy. Using Stackelberg game to model patrolling accounts for the intruder's ability to observe the patrollers' behavior and formulate an intrusion plan which exploits weak points in the patrollers' strategies. We observe in our scenario, though, that Stackelberg game strategies are overly conservative. In large scale patrolling scenarios, the intruders cannot observe patrollers strategies without being in the environment. Moreover, in dynamic decentralized situations where the patrollers are recomputing their patrolling strategies on the fly when given new information about the state of their team, the intruders have little leverage to observe and learn patrollers' strategies.

A similar line of work [11, 16] models the patrolling game with multiple intruders as a two-player game in which one patrolling player selects strategies in the patrolling robots' joint action space. Since the patrollers' joint strategy space grows exponentially in the number of agents, a common simplifying assumption is that of linearly separable utilities, which decouples the patrolling robots' strategies by considering agents which can maximize their payoffs independently of one another. In [17], the authors use this assumption to justify a treatment of the multi-patroller game in which patrolling strategies can be computed by finding a minimax strategy for each patroller individually.

For a comprehensive survey of the related work, we direct the interested reader to [5, 6].

2.3 Background

As already mentioned in the introduction, there are a number of approaches to doing multi-agent/player coordination. The simplest strategy is to use a random coordination strategy, which makes no assumption about intruder behavior or mission. Another alternative is an independent minimax strategy where each patroller can achieve at least their individual worst-case outcome without building a coalition. Finally, we have our proposed polymatrix approach which is explained in the next section.

2.3.1 Polymatrix Games

The polymatrix game formalism provides a useful abstraction for describing a network of pairwise inter-actions between players, where the nodes in the network correspond to players, and the edges represent the pairwise interactions between them. A polymatrix game $\mathcal{G} = (\mathcal{V}, \mathcal{E}, \mathcal{S}, \mathcal{U})$ is defined as [12]:

- A finite set $\mathcal{V} = 1, \ldots, n$ which correspond to vertices (players) in the interaction graph.
- A finite set $\mathcal{E}$ of edges connecting players in $\mathcal{V}$, denoting a game-interaction between the connected players.
- A set of strategies $\{\mathcal{S}\}^n$ for each player, which defines the set of strategy profiles $\mathcal{S} = \prod_{i=1}^{n} \mathcal{S}_i$.
- For each edge $(i, j) \in \mathcal{E}$, a corresponding two-player game defined by $\left(u^{ij}, u^{ji}\right)$, where $u^{ij} : \mathcal{S}_i \times \mathcal{S}_j \rightarrow \mathbb{R}$ is the payoff function for player i in the game it plays with player j. Let $\mathcal{U} = u^{ij} : (i, j) \in \mathcal{E}$ be the set of all the pairwise-payoff functions.

For each strategy profile $s \in \mathcal{S}$, the total utility to player i is given by the sum over their neighbor's utilities$u_i(s) = \sum_{j \in Nbr(i)} u^{ij}\left(s_i, s_j\right)$, where Nbr$(i) = j : (i, j) \in \mathcal{E}$ denotes the neighbors of the node i in the graph.

If for all strategy profiles $s \in \mathcal{S}$, we have $\sum_{i \in \mathcal{V}} u_i(s)$, then we say that $\mathcal{G}$ is *zero-sum*. If we have that for all $(i, j) \in \mathcal{E}$ and $s \in \mathcal{S}$, $u^{ij}\left(s_i, s_j\right) + u^{ji}\left(s_j, s_i\right) = 0$, then we say that $\mathcal{G}$ is pairwise zero-sum. Note that each player must commit to a single strategy to play in all games in which it participates.

In [12], the authors show that we can compute a Nash equilibrium strategy profile for a zero-sum polymatrix game $\mathcal{G}$ by solving the linear program.

$$\min_{\{y_i, w_i\}_{i=1}^{n}} \sum_{i \in \mathcal{V}} w_i, \text{ s.t. } w_i \geq u_i\left(s_i, y_{-i}\right), \forall i \in \mathcal{V}, \forall s_i \in \mathcal{S}_i, \text{ and } y_i \in \Delta(\mathcal{S}_i), \forall i \in \mathcal{V} \tag{2.1}$$

where $\boldsymbol{y}_{-i}$ denotes the mixed strategies of players in $\mathcal{V}\backslash i$ and $\Delta(\mathcal{S}_i)$ is the set of probability distributions over Si. They also show that the optimal objective value of (1) is zero, and that a Nash equilibrium is obtained at this value. If $\mathcal{G}$ is pairwise zero-sum, then each player has a unique payoff that it receives at any Nash equilibrium, known as its value in the game [18].

The most salient difference between the Linear Program in (1) and an independent minimax formulation is that, rather than optimize a lower bound on their own worst-case payoff, each player seeks to place an upper on the payoffs of the players with whom they compete.

2.4 Problem Formulation

We are interested in best-effort patrolling of a given area to defend it from adversarial intruders, which we will do by deploying a team of n robots. Each patrolling robot takes on one type from a set of robot types $\Lambda =$ Ground, Aerial. The intruder may appear as either a ground or aerial robot. Aerial robots have an advantage with respect to their detection range and can detect intruders at greater distances than ground units. We also consider, though, that they are vulnerable to ground units: Aerial robots are averse to coming into the detection range of ground units, since ground units may be able to attack aerial robots in a real patrolling scenario.

Let the patrolling area be approximated by a directed road network $\mathcal{R} = (\mathcal{N}, \mathcal{T})$ where $\mathcal{N} = \boldsymbol{p}_1, \ldots, \boldsymbol{p}_{|\mathcal{N}|}$ are the nodes (road junctions) and $\mathcal{T}$ is the set of possible transitions (directed road segments) between nodes. Each patroller's strategy set consists of the transitions in $\mathcal{T}$. We interpret the mixed strategies $\{\boldsymbol{x}^{(i)}\}_{i=1}^{n}$ as the probabilities that patrolling agents choose to patrol along each road segment in $\mathcal{R}$. The goal of the patrollers is to detect intruders, who attempt to go undetected while occupying road segments in $\mathcal{R}$. Patrollers also want, however, to avoid detection by intruders as much as possible. Otherwise, patrolling robots risk confrontations with adversarial units who can damage them. Thus, unlike typical adversarial patrolling, we assume that both patrollers and intruders want to avoid detection by their opponents.

Let $u_p, u_a : \mathcal{T}^{n+1} \times \Lambda \to \mathbb{R}$ represent the utility for the patrolling team and the intruder, respectively. The payoffs are dependent upon the transitions in $\mathcal{T}$ chosen by the patrollers and the intruder, as well as the robot type used by the intruder. We assume that the game played between the patrolling team and the intruder is *zero-sum* i.e. $u_p = -u_a$. We also assume that the patrollers' utilities are linearly separable [17]. This means that there exist utility functions $u^{(i)} : \mathcal{T}^2 \times \Lambda \to \mathbb{R}$, one for each patroller, such that $u_p(t_1, \ldots, t_n, t_a, \lambda_a) = \sum_{i=1}^{n} u^{(i)}(t_i, t_a, \lambda_a)$, where $\{t_i\}_{i=1}^{n}$ and t_a are the transitions in T chosen by the patrollers and intruder, respectively, and λa is the robot type chosen by the intruder.

In order to ensure that the probability of an agent transition to a particular node remains consistent, we enforce that the probability of transitioning into a node must be equal to the probability of transitioning out of the node,

$$x^{(i)} \geq 0, \sum_{t \in \mathcal{T}} x_t^{(i)} = 1, \text{ and } \sum_{t \in In(v)} x_t^{(i)} = \sum_{t \in Out(v)} x_t^{(i)}, \forall v \in \mathcal{N}; \forall i \in 1, \ldots, n \tag{2.2}$$

where $x_t^{(i)}$ denotes the tth element of the mixed-strategy vector $\boldsymbol{x}^{(i)}$, In(v) and Out(v) denote the incoming and outgoing transition for a node, respectively. Similarly, we denote by $y \in \Delta(\mathcal{T} \times \Lambda)$ the mixed strategy for the intruder in choosing the road segments to traverse as well as the intruder type. We enforce a similar constraint for a valid intruder strategy y,

$$\boldsymbol{y} \geq 0, \sum_{\lambda \in \Lambda} \sum_{t \in \mathcal{T}} y_{t,\lambda} = 1, \text{ and } \sum_{t \in \text{In}(v)} y_{t,\lambda} = \sum_{t \in \text{Out}(v)} y_{t,\lambda}, v \in \mathcal{N}, \lambda \in \Lambda \tag{2.3}$$

Formally, we model the patrolling problem as a multi-player game where patrollers try to maximize their utility by choosing appropriate mixed strategies $\{x^{(i)}\}_{i=1}^{n}$ while the intruder tries to minimize the patroller tries to minimize the same,

$$\max_{\{\boldsymbol{x}^{(i)}\}_{i=1}^{n}} \min_{\boldsymbol{y}} \sum_{i=1}^{n} u^{(i)}\left(\boldsymbol{x}^{(i)}, \boldsymbol{y}\right), \text{ s.t. (2), (3)} \tag{2.4}$$

2.5 Utility Model

The patroller's utility is dependent on a detection event. A patrolling robot at position $\boldsymbol{p}_d$ is said to detect an attacker at position $\boldsymbol{p}_a$ when the attacker comes within the detection range rd of the patrolling robot.

$$\delta\left(\boldsymbol{p}_d \rightarrow \boldsymbol{p}_a\right) = \begin{cases} 1 \text{ if } \|\boldsymbol{p}_d - \boldsymbol{p}_a\|_2 < r_d \text{ and } \boldsymbol{p}_a \text{ is in line of sight of } \boldsymbol{p}_d \\ 0 \quad \text{otherwise} \end{cases} \tag{2.5}$$

The detection is said to be mutual when both $\boldsymbol{p}_d$ and $\boldsymbol{p}_a$ are within each other's detection range $\delta\left(\boldsymbol{p}_d \rightarrow \boldsymbol{p}_a\right) = \delta\left(\boldsymbol{p}_a \rightarrow \boldsymbol{p}_d\right) = 1$.

We model two kinds of utility, occupation utility $u_{\text{occ}}(\boldsymbol{p}, \lambda) \in \mathbb{R}^+$ and detection utility $u_{\text{det}} \in \mathbb{R}^+$. In the absence of a detection event, the patrollers get a utility of $-u_{\text{occ}}\left(\boldsymbol{p}_a, \lambda_a\right)$ while the intruder gets $u_{\text{occ}}\left(\boldsymbol{p}_a, \lambda_a\right)$. Whenever, a detection event occurs the detection utility is determined by three factors, (1) whether the detection is mutual (both players detect each other) or asymmetric (2) robot type (ground or aerial) and (3) team (patroller or intruder). In case of a mutual detection, the Ground robot has the advantage against the Aerial robot, hence the patroller utility is positive if its robot type is Ground.

$$u^{(p)}_{\text{mutdet}}(\lambda_d, \lambda_a) = \begin{cases} u_{\text{det}} if \lambda_d = \text{Ground}, \lambda_a = \text{Aerial} \\ -u_{\text{det}} if \lambda_d = \text{Aerial}, \lambda_a = \text{Ground} \\ 0 \quad \text{otherwise} \end{cases} \tag{2.6}$$

On the other hand, in case of an asymmetric detection, the detector gets the u_{det}, while the detected gets $-u_{\text{det}}$. When the detector is an intruder the patroller gets $u_{\text{det}} - u_{\text{occ}}(\boldsymbol{p}_a, \lambda_a)$

$$\mathrm{u}^{(\mathrm{p})}_{\text{pos}}(\boldsymbol{p}_\mathrm{d}, \boldsymbol{p}_\mathrm{a}, \lambda_a) \begin{cases} u^{(d)}_{\text{mutdet}}(\lambda_d, \lambda_a) \text{ if } \delta(\boldsymbol{p}_d \to \boldsymbol{p}_a) \text{ and } \delta(\boldsymbol{p}_a \to \boldsymbol{p}_d) \\ u_{\text{det}} - u_{\text{occ}}(p_a, \lambda_a) \text{ if } \delta(\boldsymbol{p}_a \to \boldsymbol{p}_d) \text{ and not } \delta(\boldsymbol{p}_d \to \boldsymbol{p}_a) \\ u_{\text{det}} \text{ if } \delta(\boldsymbol{p}_d \to \boldsymbol{p}_a) \text{ and not } \delta(\boldsymbol{p}_a \to \boldsymbol{p}_d) \end{cases} \tag{2.7}$$

and $-u_{\text{occ}}(\boldsymbol{p}_a, \lambda_a)$otherwise.

Now that we have the utilities for the positions of the robots, we can define utilities for transitions over road segments. Assuming a constant velocity model, the position is parameterized over the road segment using normalized time $\rho \in (0, 1)$. The position of the patroller at any time ρ is given by $\boldsymbol{p}_d(\rho) = (1 - \rho) \cdot \text{start}(t_d) + \rho \cdot \text{end}(t_d)$. The utility over road segments $t_d, t_a \in \mathcal{T}$ is computed by integrating the positional utility over the linear road segment.

$$u^{(d)}(t_d, t_a, \lambda_a) = \int_0^1 u^{(d)}_{\text{pos}}(\boldsymbol{p}_d(\rho), \boldsymbol{p}_a(\rho), \lambda_a) d\rho. \tag{2.8}$$

Note that utility computation has to be done only once for a static map and a particular robot type.

2.6 Game Theoretic Patrolling

We consider two approaches, each with a different set of assumptions to address the proposed problem (1) Independent Minimax Approach and (2) Polymatrix Approach.

2.6.1 Independent Minimax Approach

A naïve approach to solving the multi-patroller problem would be to play a minimax game in the joint strategy space of the patrollers. This approach, however, is not feasible with even a small number of patrollers. We follow an alternative approach where each patroller is modeled as an independent player that plays a minimax game against a single intruder. This approach scales linearly with the number of intruders. Moreover, each patroller-intruder game becomes a two-player zero-sum game which can be efficiently solved with a linear program:

$$\max_{\{x_i\}_{i=1}^n} z_i, \text{ s.t.} z_i 1 \leq \boldsymbol{x}^{(i)\top} \boldsymbol{U}^{(i)}, \text{ and (2)}, \tag{2.9}$$

where $z_i \in \mathbb{R}$ is the lower bound on the ith patroller's utility given the possible actions taken by the intruder. Here, $\boldsymbol{U}^{(i)} = [[u^{(i)}(t_i, t_a, \lambda)]_{t_i \in \mathcal{T}}]_{t_a \in \mathcal{T}, \lambda \in \Lambda}$ is the utility matrix for the ith patroller.

2.6.2 Polymatrix Approach

We propose the use of polymatrix games [12] for the task of adversarial patrolling. We instantiate the polymatrix game $\mathcal{G} = (\mathcal{V}, \mathcal{E}, \mathcal{S}, \mathcal{U})$ for adversarial patrolling as follows.

- Each robot, the n patrollers and an intruder are modeled as separate players, i.e. $\mathcal{V} = 1, \ldots, n+1$, where $n+1$ is the intruder.
- We define a set of pairwise zero-sum games between each patroller and the intruder as $\mathcal{E} = \{(k, n+1)\}_{k=1}^n$.
- The mixed strategy of each patroller is a probability distribution over all edges of the road network $\mathcal{R}$, i.e. $\mathcal{S}_i = \mathcal{T}$ for all $i \in 1, \ldots, n$. The intruder also chooses its robot type, i.e. $\mathcal{S}_{n+1} = \mathcal{T} \times \Lambda$.
- The utilities are as modeled in Sect. 2.5, so that a set of utility matrices $\boldsymbol{U}^{(i)} = [[u^{(i)}(t_i, t_a, \lambda)]_{t_i=1}^{\mathcal{T}}]_{t_a=1, \lambda \in \Lambda}^{\mathcal{T}}$ define the payoff for each patroller.

Such a polymatrix formulation can be solved to obtain the Nash equilibrium as a linear program:

$$\min_{\{x\}_{i=1}^n, y} \sum_{i=1}^n z_i + w, \ s.t. w 1 \geq -\sum_{i=1}^n \boldsymbol{x}^{(i)T} \boldsymbol{U}^{(i)}, z_i 1^T$$
$$\geq \boldsymbol{U}^{(i)} \boldsymbol{y} \forall i \in 1, \ldots, \text{n, (2), and(3)} \tag{2.10}$$

where z_i is the upper bound on the ith patroller's utility and w is the upper bound on the intruder's utility. Even though solving this problem finds an approximate, and not an exact, Nash equilibrium, due to the extra constraints on movement, we show in Sect. 2.7 that our method still performs well despite the intruder being allowed to exploit the gap in utility caused by the error.

2.7 Experimental Evaluation

In this section, we compare the performance of our approach using polymatrix games to patrolling strategies obtained using independent minimax strategies and uniformly random patrolling strategies.

Intruder strategy optimization. We are interested in comparing the worst case scenario for the pa- trolling team given that they follow strategies prescribed by each of the three methods. Hence, to evaluate the performance of each of the three patrolling approaches, we compute the intruder's strategy by finding a best response given the patrollers' strategies. Specifically, given patroller strategies $\boldsymbol{x}^{(1)}, \ldots, \boldsymbol{x}^{(n)}$, we solve the problem,

$$\max_{\mathrm{y}} \sum_{i=1}^{n} \boldsymbol{x}^{(i)\top} \boldsymbol{U}^{(i)} \boldsymbol{y} \text{s.t. (3)} \tag{2.11}$$

2.7.1 Experimental Setup

We use 100 randomly generated gridworld environments to validate our computed patrolling strategies, of which Fig. 2.2 shows five examples. The environments were generated by starting with a 10×10 grid of vertices and randomly removing vertices with a probability of 0.35. After that, vertices not connected to the vertex at the point (0, 0) were also removed to ensure that all the generated graphs are connected. In our experiments, we interpret white space between vertices as buildings through which robots can not detect each other. We assume that aerial robots fly below the tops of buildings in order to avoid making themselves too highly visible to adversaries. Hence, all robots can detect other robots only if they are located at vertices which are connected to their current location by an unbroken, straight path and are within their sensing range. In our simulations, ground robots' sensing range extends to adjacent vertices, while aerial robots may detect other robots up to 3 edges away.

We run each simulation for 20 time steps, where patrollers and intruders spawn at locations according to estimated strategies. At every time step, both the patrollers and the intruder sample a nearby edge to move from their respective optimal strategies. We evaluate each method based on the proportion of intruders detected, the number of times a patroller is detected by an intruder, and the average time taken to detect

Fig. 2.2 Examples from randomly generated gridworld environments

each intruder. We present results for two different patroller teams, each computing strategies using $u_{\text{occ}} = 1$ and $u_{\text{occ}} = 2$. By varying the value of u_{occ}, we compare the results when facing an intruder which either has no preferences between occupation and detecting patrollers, or which prioritizes occupying territory without detection, respectively. The first team consists of two ground robots and one aerial robot, while the second team consists of one ground robot and two aerial robots. For each of the 100 environments, we run 10,000 simulations for both values of u_{occ}, each with a duration of 20 time steps. In order to reduce the time required to compute the utility matrix, we test for detection between two robots only at vertices in the road network rather than integrating over the road segment (8).

2.7.2 Executing Computed Strategies

At the beginning of each simulation, the ith patrolling robot is initialized by sampling an edge from its mixed strategy $\boldsymbol{x}^{(i)}$ and placing the patroller on the terminal vertex of that edge. At each timestep of the simulation the ith patroller chooses a transition in Out(k) by sampling from the distribution defined by $\mathbb{P} = \frac{x_t^{(i)}}{\sum_{t \in \text{Out}(k)} x_t^{(i)}}$.

At any time during the simulation, a new intruder has a probability $p = 0.3$ of entering the environment. New intruders are initialized by first drawing from the distribution $\mathbb{P}(\lambda) = \sum_{t \in T} y(t, \lambda)$ to determine the robot type as which they will instantiate themselves. The intruder's position is then initialized in a way analogous to the patrollers. An intruder may not change its type after entering the environment. Once an intruder is detected by a patroller, that intruder is considered to be neutralized and has no effect on the rest of the simulation (cannot be detected again and cannot detect patrollers).

2.7.3 Simulation Results

Quantitative simulation results are shown in Fig. 2.3 for the two patrolling team configurations and the parameters $u_{\text{occ}} = 1$ and $u_{\text{occ}} = 2$.

We see that our approach using polymatrix games is superior to the baseline method with respect to the proportion of intruders detected in each simulation, regardless of team composition and value of u_{occ}. This demonstrates that the improvement in the rate of intruder detection our method provides is robust to changes in model parameters and team composition. With respect to the time for which intruders were allowed to occupy the road network without detection, the polymatrix and individual minimax approaches are the same, but both methods perform better than random patrolling strategies on this metric. Recall also that we measure the time-to-detection based on the intruders that were detected in the first place. This means that robots

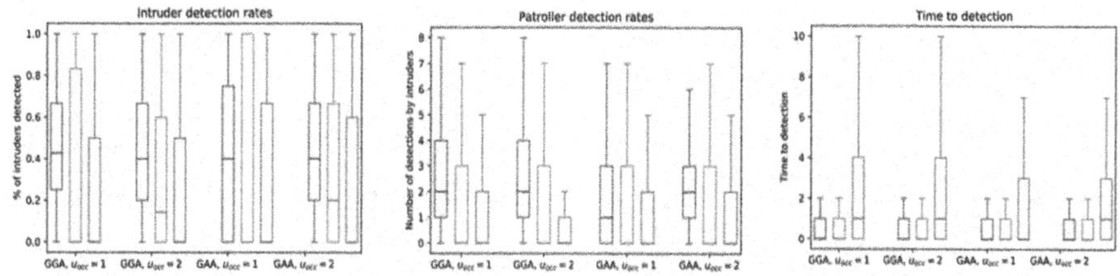

Fig. 2.3 Simulation results for different team compositions and values of u_{occ}. GGA denotes a patroller team composed of two ground and one aerial robot, while GAA denotes a team of one ground and two aerial robots. The results for the polymatrix, independent minimax, and random patrolling strategies are blue, orange, and green, respectively. We see that our approach using polymatrix games is superior to the baseline method with respect to the proportion of intruders detected in each simulation, although with a tradeoff of marginal increase in patrollers detected

using polymatrix strategies are able to detect more intruders, just as quickly as when they use their independent minimax strategies.

We see also, however, that patrolling agents using our polymatrix strategies are more likely to be detected by intruders. This reflects the inherent tradeoff between detecting intruders and avoiding detection by them. We observe that patrollers who detect more intruders are more likely to come within detection range of those intruders. Our results also show, however, that this is a reasonable tradeoff to make: the increase in performance in terms of the proportion of intruders detected when patrollers use their polymatrix strategies greatly outweighs the increase in their likelihoods of being detected.

2.8 Application to a Realistic Environment

In this section, we show how to adapt our approach to the problem of multi-robot patrolling in a large-scale, realistic simulated environment. We use a custom Unity based simulator in which robots can be spawned and moved around with realistic physics and graphics (See Fig. 2.4). We improve upon the realism of Sect. 2.5 by incorporating different movement speeds for each robot type (aerial robots move more quickly than ground robots). We also use a more sophisticated model of visibility between points and consideration of the terrain complexity, as well as a measure of a robot's range of visibility at each position, along each road segment. In this section, we consider an intruder who wants to occupy road segments along which it has high visibility but which have low terrain complexity. In other words, intruders seek to maximize the extent to which they can survey the surrounding area while minimizing the cost of traveling through the environment.

We assume in this section that we are given maps assigning terrain complexity and visibility scores to each point in the environment for each robot type. We compute the occupation utility for the intruder at a point $\boldsymbol{p}$ by $u_{\text{occ}}(\boldsymbol{p}, \lambda) = u_{\text{vis}}(\boldsymbol{p}, \lambda) - u_{\text{terr}}(\boldsymbol{p}, \lambda)$, where u_{vis} and u_{terr} are mappings assigning a visibility score and terrain complexity

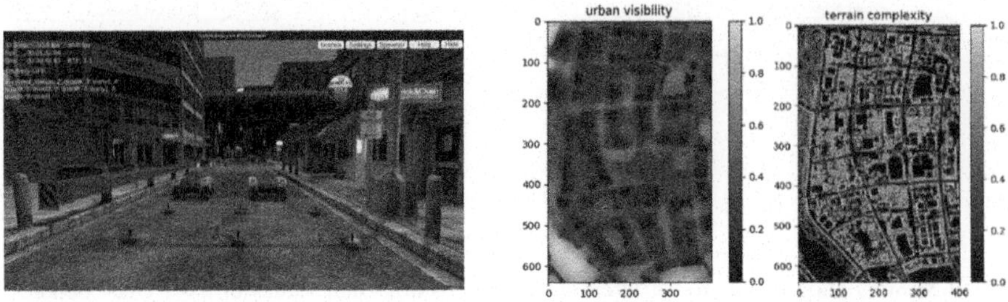

Fig. 2.4 From left to right: (1) First person view of the unity based realistic simulator in which the patrolling experiments were done. (2) Visibility heatmap for the patrolling region used as $u_{\text{vis}}(\boldsymbol{p}, \lambda)$. (3) Terrain complexity heatmap for the patrolling environment in the simulator used as $u_{\text{terr}}(\boldsymbol{p}, \lambda)$

to each point and robot type (See Fig. 2.4). We then normalize the occupation utilities to the range [0, 1].

2.8.1 Qualitative Analysis of Patrolling Strategies

Figure 2.5 shows visualizations of the strategies chosen by the patrollers and the intrusion plan chosen by the intruder under the Nash equilibrium strategy profile. Note that the two ground units choose the same strategies, so we visualize a strategy for only one of them. The edge colors indicate the probability that a robot chooses to move along the corresponding road segment in either direction. The vertex colors indicate the probability of staying in the same location. The intruder strategies shown visualize the strategy used by the intruder when choosing to appear as either a ground or aerial robot.

Here, the intruder chooses to appear as an aerial robot with a probability of 0.89. Choosing to send mostly aerial robots in this environment makes sense, since they have greater visibility everywhere in the environment than ground robots do and can move at greater speeds. This allows the intruder to maximize the utility it receives from occupying the road network without being detected. Additionally, using ground robots is less advantageous for the intruder since the patrolling team has more aerial robots than ground robots: Recall that the primary advantage of ground robots is that they receive a high payoff by getting sufficiently close to aerial robots. Finally, aerial robots can more easily obtain high occupation utilities, since they have low or no terrain complexity everywhere and high visibility relative to ground robots.

A qualitative analysis of the strategies shown in Fig. 2.5 shows that they exhibit many intuitive features. We see that patrolling ground units are likely to patrol along road segments where they can detect intruding aerial robots. We see also a division of labor between patrolling ground and aerial robots: Ground robots patrol in areas along the perimeter of the road network, while aerial robots patrol the interior of the urban environment. This is because aerial units can fly over and see above smaller

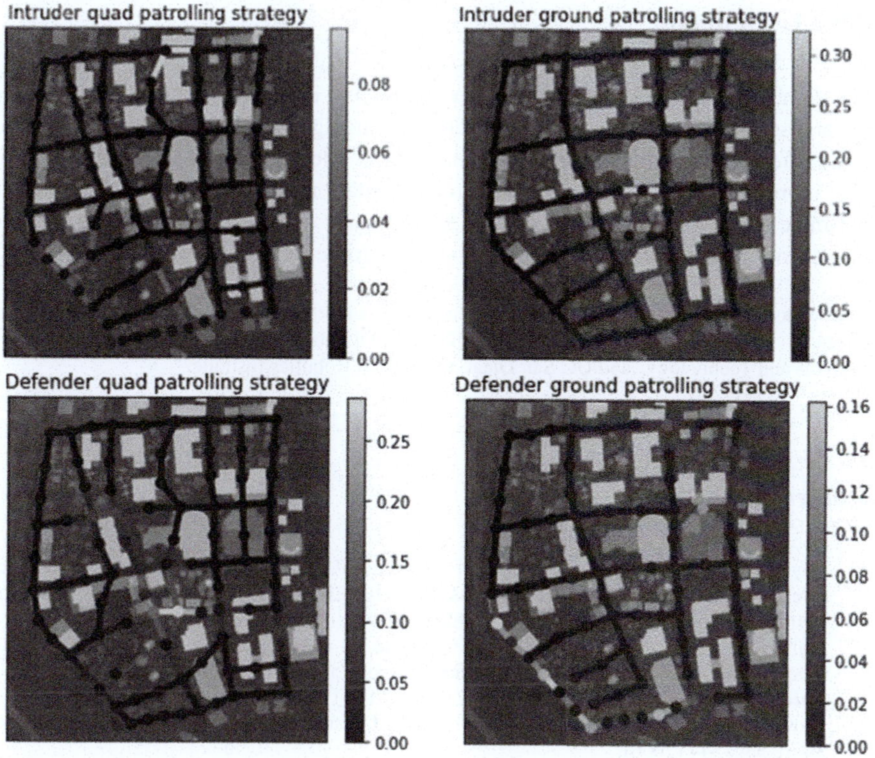

Fig. 2.5 In order from left to right: patrolling aerial robot strategy, patrolling ground robot strategy, intruding aerial robot strategy, intruding ground robot strategy

buildings, making them more suitable to detect intruders who attempt to hide between buildings.

Intruding aerial robots, on the other hand, attempt to occupy the perimeter of the environment. One reason for this is that they can obtain the most utility for occupying the bottom-left region of the environment. This is a highly valuable area of the environment for intruding aerial robots as it affords a high level of visibility. As for intruding ground robots, their role on the intruding team is oriented toward detection. They choose to occupy areas where they are more likely to detect patrolling aerial robots.

Given that we can compute interpretable and reasonable patrolling behavior in this environment, we see that our method is applicable to a realistic environment, and therefore that our approach is suitable for computing patrolling strategies in real-world scenarios. Furthermore, we have shown that our game- playing model is easily extensible: We can easily incorporate sophisticated models of visibility and terrain analysis to improve the realism of the generated strategies.

2.9 Conclusion

We demonstrate the use of polymatrix games for effective adversarial patrolling in large-scale environments using heterogeneous agents. The proposed method is more effective in capturing intruders than independent minimax formulation but is also more amenable to decentralized implementation as compared to methods that optimize in the joint strategy space. In future work, we will implement this method on a team of decentralized robots that is robust to network failures.

Acknowledgements We gratefully acknowledge the support of this research from the Army Research Lab—Collaborative Research Alliance—"Distributed and Collaborative Intelligent Systems and Technology" and UC San Diego Contextual Robotics Institute.

References

1. Couceiro, M.S., Portugal, D., Ferreira, J.F., Rocha, R.P.: Semfire: towards a new generation of forestry maintenance multi-robot systems. In: 2019 IEEE/SICE International Symposium on System Integration (SII), pp. 270–276. IEEE (2019)
2. Marino, A., Parker, L.E., Antonelli, G., Caccavale, F.: A decentralized architecture for multi-robot systems based on the null-space-behavioral control with application to multi-robot border patrolling. J. Intell. Robot Syst. **71**(3), 423–444 (2013)
3. Brown, G., Carlyle, M., Salmerón, J., Wood, K.: Defending critical infrastructure. Interfaces **36**(6), 530–544 (2006)
4. Zoss, B.M., Mateo, D., Kuan, Y.K., Tokić, G., Chamanbaz, M., Goh, L., Vallegra, F., Bouffanais, R., Yue, D.K.: Distributed system of autonomous buoys for scalable deployment and monitoring of large waterbodies. Auton. Robot. **42**(8), 1669–1689 (2018)
5. Portugal, D., Rocha, R.: A survey on multi-robot patrolling algorithms. In: Doctoral Conference on Computing, Electrical and Industrial Systems, pp. 139–146. Springer (2011)
6. Huang, L., Zhou, M., Hao, K., Hou, E.: A survey of multi-robot regular and adversarial patrolling. IEEE/CAA J. Autom. Sinica **6**(4), 894–903 (2019)
7. Shishika, D., Kumar, V.: A review of multi agent perimeter defense games. In: International Conference on Decision and Game Theory for Security, pp. 472–485. Springer (2020)
8. Talebi, S., Simaan, M.A., Qu, Z.: Cooperative, non-cooperative and greedy pursuers strategies in multi-player pursuit-evasion games. In: 2017 IEEE Conference on Control Technology and Applications (CCTA), pp. 2049–2056. IEEE (2017)
9. Vidal, R., Shakernia, O., Kim, H.J., Shim, D.H., Sastry, S.: Probabilistic pursuit-evasion games: theory, implementation, and experimental evaluation. IEEE Trans. Robot. Autom. **18**(5), 662–669 (2002)
10. Frost, J., Stone, L.: Review of search theory: advances and applications to search and rescue decision support us coast guard research and development center rep. Tech. Rep. CG-D-15-01 (2001)
11. Delle Fave, F.M., Jiang, A.X., Yin, Z., Zhang, C., Tambe, M., Kraus, S., Sullivan, J.P.: Game-theoretic patrolling with dynamic execution uncertainty and a case study on a real transit system. J. Artif. Intell. Res. **50**, 321–367 (2014)
12. Cai, Y., Candogan, O., Daskalakis, C., Papadimitriou, C.: Zero-sum polymatrix games: a generalization of minmax. Math. Oper. Res. **41**, 648–655 (2016)
13. Paruchuri, P., Pearce, J.P., Tambe, M. Ordonez, F., Kraus, S.: An efficient heuristic approach for security against multiple adversaries. In: Proceedings of the 6th International Joint Conference on Autonomous Agents and Multiagent Systems, pp. 1–8 (2007)

14. Paruchuri, P, Pearce, J.P., Marecki, J., Tambe, M., Ordonez, F., Kraus, S.: Playing games for security: an efficient exact algorithm for solving bayesian stackelberg games. In: AAMAS (2008)
15. Vorobeychik, Y., An, B., Tambe, M., Singh, S.: Computing solutions in infinite-horizon discounted adversarial patrolling games. In: Proceedings of the International Conference on Automated Planning and Scheduling, vol. 24 (2014)
16. Alam, T., Edwards, M., Bobadilla, L., Shell, D.: Distributed multi-robot area patrolling in adversarial environments. In: International Workshop on Robotic Sensor Networks (2015)
17. Jiang, A.X., Yin, Z., Zhang, C., Tambe, M., Kraus, S.: Game-theoretic randomization for security patrolling with dynamic execution uncertainty. In: Proceedings of the 2013 International Conference on Autonomous Agents and Multi-Agent Systems, pp. 207–214 (2013)
18. Cai, Y., Daskalakis, C.: On minmax theorems for multiplayer games. In: Proceedings of the Twenty-Second Annual ACM-SIAM Symposium on Discrete Algorithms, pp. 217–234. Society for Industrial and Applied Mathematics, Philadelphia, PA (2011)

Chapter 3
Geospatial Big Data Management Testing Using Open Source Technology

Solechoel Arifin, Florence Elfriede Sinthauli Silalahi, Mugi Prayitno, Nur Kholis Majid, Fahmi Amhar, and Herjuno Gularso

Abstract Geospatial big data multiply since the world becomes more mobile and because much of the data in the world can be geo-referenced. Especially in a country plan, it can be used along with statistical and financial data. Pilot big data projects start with understanding the technology in handling data at scale and learning to make Hadoop work the way the user desires. In this research, we want to test geospatial big data management using YAVA and Hadoop GRID (HGRID) as improved modification of Hadoop owned by PT 247 with data from landslide events in Bogor, West Java, and salt mapping identification in East Java (raster and vector data). YAVA Hadoop is a compilation of Big Data platforms where the platforms are open source that leverages the power of the Apache Hadoop ecosystem and is designed to help accelerate Hadoop deployments and their ecosystem in Indonesia. Meanwhile, HGrid is an open-source software framework for storing data and running applications on a locally made group of commodity hardware or a big data engineering tool (ETL Tool). It is a GUI tool that helps the user easily design data engineering workflows to process big data automatically. We found that YAVA is user-friendly in managing geospatial big data, and its stack component includes HDFS, MapReduce, Yarn, Hbase, and Hive. However, YAVA does not provide a public online repository, so a local repository must be created. In our testing, the spatial menu in HGRID is helpful with limitations in visual-spatial view, and the output report is informative enough.

Keywords Big data · Geospatial · Hadoop · MapReduce · YAVA · HGRID

S. Arifin · M. Prayitno · N. K. Majid
Solusi247, Jl. Prof. Dr. Satrio Kav 6 South Jakarta, Jakarta, Indonesia

F. E. S. Silalahi (✉) · F. Amhar
National Research and Innovation Agency, Jl. Raya Bogor Km. 47, Cibinong, West Java, Indonesia
e-mail: flor006@brin.go.id

H. Gularso
Geospatial Information Agency, Jl. Raya Bogor Km. 46, Cibinong, West Java, Indonesia

M. A. Laribi et al. (eds.), *Advances in Automation, Mechanical and Design Engineering*, Mechanisms and Machine Science 121,
https://doi.org/10.1007/978-3-031-09909-0_3

3.1 Introduction

Nowadays, Geospatial data is publicly disseminated and originated by public and private organizations with the fundamental build once and share or use many times. It has multiplied since the world has become more mobile, and big data platforms have been developed so that users can implement big data analytics for modelling and simulation of geospatially enabled content these days [1, 2]. Geospatial big data comes from the many different types of field measurement, sensors, observation systems, and social networks because much of the data in the world can be georeferenced. Geospatial big data needs to handle and visualize other types of massive data sets, such as extensive collections of high-resolution remote-sensing data and radar data [3].

Big data is one of the concerns, especially concerning the One Data policy in Indonesia. Inevitably, the more rapid development of technology today makes more data is generated each day. There are countless numbers of distributed data. The rise of big data is triggered by technology growth, internet adoption, people's behavior, digitalization, and competition. Geospatial Big Data represents big data in the form of layers and spatial attributes. Data and information generated and reprocessed from service systems and equipment sensors require a reliable storage system in terms of operational management, performance, ease of addition, and interoperability with other systems [4].

This is supported by regulations related to the principle of open data in the dissemination or dissemination of public data, namely Law No. 14 of 2008 concerning Public Information Disclosure. Presidential Regulation No. 95 of 2018 concerning the Electronic-Based Government System (SPBE) article 41 paragraph 1, namely every ministry/agency/local government must implement security on SPBE [5]. The regulation of One Indonesian Data aims to (1) provide implementation reference and guidelines for Central Agencies and Regional Agencies in the context of implementing data management to support development planning, implementation, evaluation, and control; (2) realizing the availability of data that is accurate, up-to-date, integrated, accountable, and easily accessible and shared between Central Agencies and Regional Agencies as the basis for planning, implementing, evaluating, and controlling development; (3) encourage data openness and transparency to create data-based development planning and policy formulation; and (4) support the national statistical system of laws and regulations, through compliance with data standards, metadata, data interoperability, using reference codes and master data. Data interoperability is the ability of data to be shared between interacting electronic systems. Geospatial Data and Information are included in the data classification in question. They are the initial interoperability target in implementing the Presidential Regulation related to One Indonesian Data and statistical and financial data [6].

Big data is considered to be able to support the implementation of Geocoded Address, both in the process of taking input text, such as an address or place name, and returning the latitude/longitude location on the earth's surface for that place (geocoding), or converting geographic coordinates into a location description, usually a place name. or addressable locations (reverse geocoding). Geospatial data has always been big data because it can be characterized by (1) the volume (increasing volume of the real-time sensor), (2) the variety (complex structures such as structured and unstructured data, digital map, imagery data, raster, and vector data), (3) the velocity (continuous streaming of sensor observations of CORS and tidal), (4) the veracity (the level of accuracy varies depending on data sources), and (5) visualization (for identifying dominant patterns and relationships that emerge from extensive data analysis) [2–4].

Most of the big data components are open source, widely used today, including Apache [2, 7, 8]. We can download the code, use and modify it freely, require adequate human resources with many choices. However, parallel processing is complex, such as job balancing, hardware failure, and failover, and most analysis tasks need to combine the data. In this research, we want to test geospatial extensive data management using YAVA and Hadoop GRID (HGRID) as improved-modification of Hadoop owned by PT 247, all in one data management platform programming and scripting using Java, Python, Scala, R, SQL with HGrid247 (a visual designer). Our objectives are to introduce YAVA and Hadoop GRID (HGRID) and their example in managing geospatial data more user-friendly.

3.2 Methods

Pilot big data projects start with understanding the technology in handling data at scale and learning to make Hadoop work the way the user desires. On the IT efficiency side, improving the ability to transform, analyze, and find the most appropriate tools to acquire more data and build applications. It involves getting data and analytics into the hands of as many people as possible. The data lake and the enterprise data warehouse start to work in union. Hadoop involves improving the ability to transform and analyze data, finding the most appropriate tools to the skillset, acquiring more data, and building applications. Then, there is a need to enhance enterprise capabilities are added to the data lake as a core competence.

RDBMS or a Relational Database Management System can be used with other stored datasets, and it provides a visual representation of the data that stores in tables. However, social media or big data implements have a better ability to accommodate irregularities of data structure and schema. Even though the concept of traditional RDBMS is still applied, RDBMS is not always the best solution for all situations as it cannot meet the increasing growth of unstructured data. NoSQL database can provide comparable as another approach because it can construct a similar classic relational database, such as classic user/password authentication to provide security, collections to organize data, and indexes to improve performance. NoSQL databases are becoming more widespread [9]. However, for pure analytical applications, Hadoop is often more suitable [10].

Hadoop is an open-source software framework for managing data processing (an enormous processing power), storage (the massive storage for any data), and distributing large data sets and analyzing jobs across nodes in a computing cluster, then breaking it down into smaller workloads that can be run in parallel for big data applications. Hadoop implements both a programming model and a big data processing engine to parallel large data sets called MapReduce [11, 12]. The applications are grouped into several small fragments of work to process and generate large data sets. Furthermore, it provides a distributed file system (HDFS) that stores data on computer nodes. Both MapReduce and HDFS are designed so sophisticatedly that the framework automatically handles possible nodes. Mapreduce can enrich referencing data. Apache Hadoop-related projects are Apache Pig, Apache Hive, Apache HBase, Apache ZooKeeper, Apache Cassandra, and many others [13].

Related current work in geospatial big data handling with high performance computing by Li [7] found the access patterns of a geospatial data partition (or block) are strongly linked to its neighboring partitions, co-locating the partitions that are spatially close with each other to a same computer node often improves data access efficiency in Hadoop Distributed File System (HDFS). It is evident that the interweaving of geospatial big data, cloud computing, fog computing, and artificial intelligence is driving and reshaping geospatial data science [7].

YAVA Hadoop is a compilation of Big Data platforms where the platforms are 100% open source. It is a combination of the power of the Apache Hadoop ecosystem and ease of use to accelerate the adoption of Hadoop implementation. YAVA Hadoop leverages the power of the Apache Hadoop ecosystem and is designed to help accelerate Hadoop deployments and their ecosystem in Indonesia (Fig. 3.1). The archival and storage are used to retain years of data and its intermediate format. Transformation usage as map inputs and outputs, where needed, turn unstructured data into structured at runtime (Fig. 3.2 and 3.3). For analysis, the function is to explore data in place and execute arbitrary code.

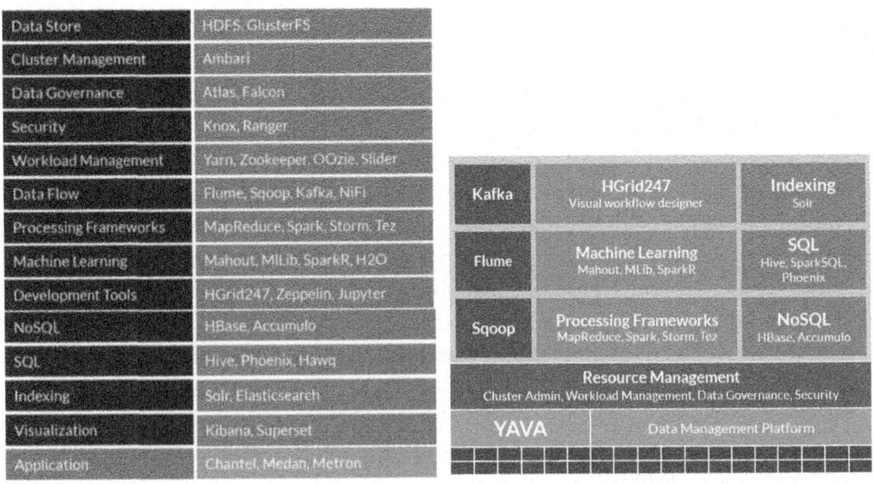

Fig. 3.1 Yava component and its data management platform

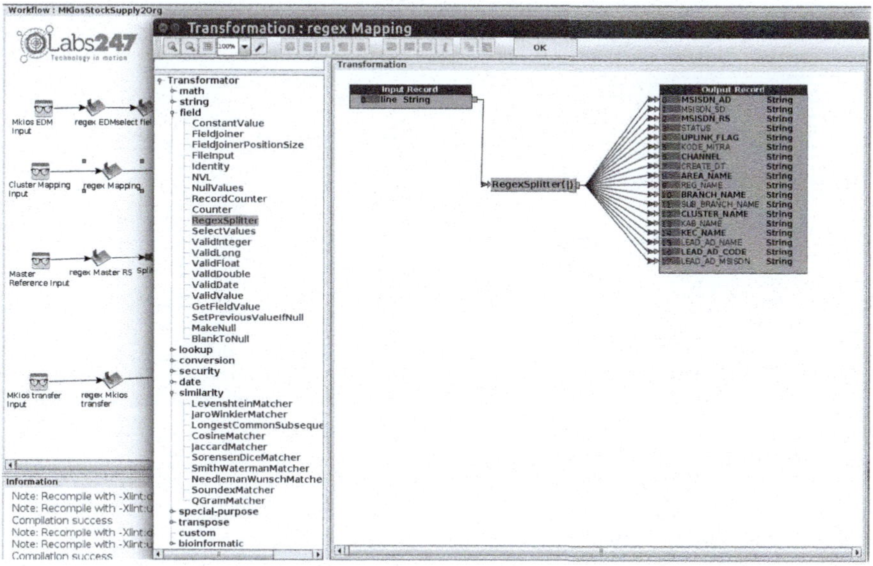

Fig. 3.2 Transformation function of HGrid

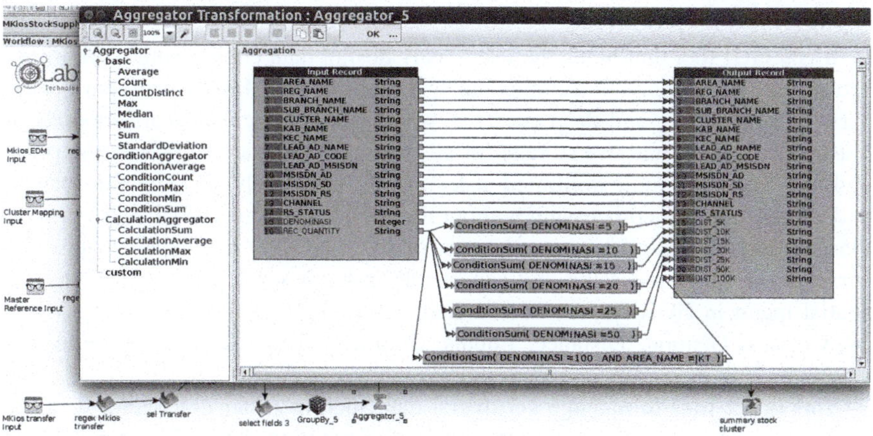

Fig. 3.3 Aggregator transformation of HGrid

A workflow in HGrid's data cleansing diagram shows the flow of processing datastream from source to sink/destination, such as the collection of source, assembly, and sink/destination. A valid Workflow must meet the following conditions, i.e., for every source, there is a path to an assembly, for every assembly, there is the path(s) from source and path(s) to sink, and for every sink, there is a path from an assembly.

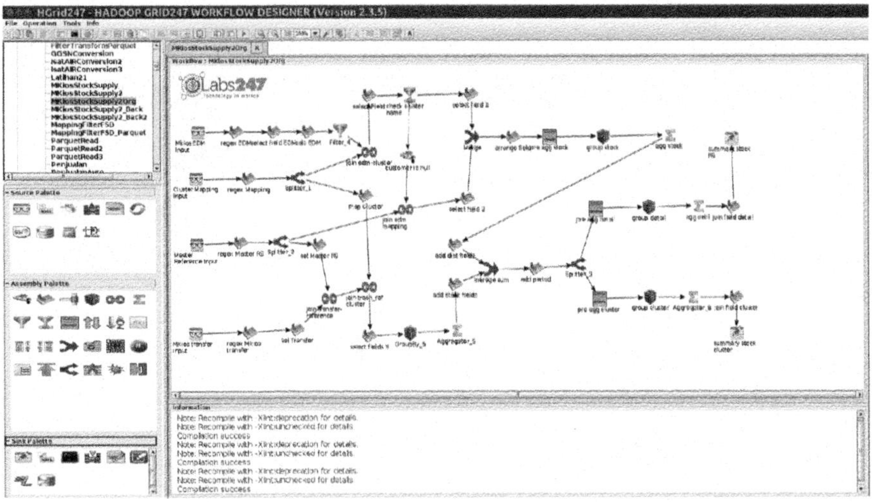

Fig. 3.4 Main window of HGrid

Hadoop Grid or HGrid is an open-source software framework for storing data and running applications on a locally made group of commodity hardware or a big data engineering tool (ETL Tool).

It is a GUI tool that helps the user easily design data engineering workflows to process big data (Fig. 3.4) automatically. It replaces the MapReduce/Spark coding activity using a drag-and-drop interface, which can be executed in a Hadoop/Spark cluster. It provides massive storage for all types of data, massive processing power, and the ability to handle unlimited tasks or jobs indefinitely. HGRID can speed up and simplify big data processing without the need for coding. HGRID also provides massive storage for all types and types of data. HGRID has excellent power in carrying out data processing and can handle tasks or work without stopping. The spatial menu in HGRID, i.e., Create Geometry, Area, Contains, Covers, Covered by, Crosses, Disjoint, Distance, Equals, Geometry Type, Intersection, Intersects, Overlaps, Union, Within, and Centroid.

This paper tries to manage geospatial data such as raster and vector data (i.e., points, lines, polygons, and other descriptive information about a location (.shp)) in our local server. GeoTIFF is one geospatial data format based on the TIFF format with georeferencing information embedded and is widely used in NASA Earth science data systems and widespread worldwide. It has TIFF tags including the following raster metadata such as spatial extent, coordinate reference system, resolution in raster format (composed of pixels), mainly as a distribution format for Digital Elevation Model (DEM), data satellite, or aerial photography imagery, and Digital Ortho Quadrangle data [14, 15].

3.3 Results

Data must be easy to access. Data governance is also needed so that there is transparent governance to ensure collection, processing, and distribution. Furthermore, there is a need for geospatial data literacy for decision-making purposes so that it is based on information from data processing. Technology enables stakeholders to collects, analyze, store, and disseminates geographic information for use by different entities to facilitate informed and smooth decision making.

We used data from landslide events in Bogor, West Java, and salt mapping identification in East Java in our testing. First, we tried HGrid247 to process vector data (IG_GARAM data). This data has properties GEOMETRY-OBJECT, ID, PENGGUNAAN (Purposes), STATUS, JENIS (Type), FUNGSI (Function), DESA (Villages), KECAMATAN (Subdistrict), KABUPATEN (Regency), PROVINSI (Province), and LUAS_HA (Unit of the area in Hectare). In this testing, we will calculate the distance between all records with ID 265,745 (data: Tambak Garam, Rakyat, Murni, Peminihan, Sreseh, Sampang, Jawa Timur).

In the first step, we set the workflow and give the class name as ShpProcess. Next, we set the parameter custom input format and get the feature of SHP file, using transformation tools on Output Record, by adding fields OBJECT, ID, PENGGUNAAN, STATUS, JENIS, FUNGSI, DESA, KECAMATAN, KABUPATEN, PROVINSI, and LUAS_HA. Then, drag and drop transformation function by drawing a link from field line of input record to RegexSpliter, and draw links from RegexSpliter to all fields of output record as seen in Fig. 3.5.

We continue by splitting all data with ID 265,745 and other data using SplitFilter, then processing match and unmatched data. A link from ConstantValue to field GROUP was added, and parameter '1' to ConstantValue parameter was set. We continue to create reference join between unmatched and match data to calculate the distance by adding the field JARAK (distance) on OutputRecord, so the distance between OBJECT with ID = 265,745 and others OBJECTS can be recorded. The property of Transformator was set to identity all fields from Input Record to Output Record, i.e., the transformation-function distance of Spatial Transforms and select

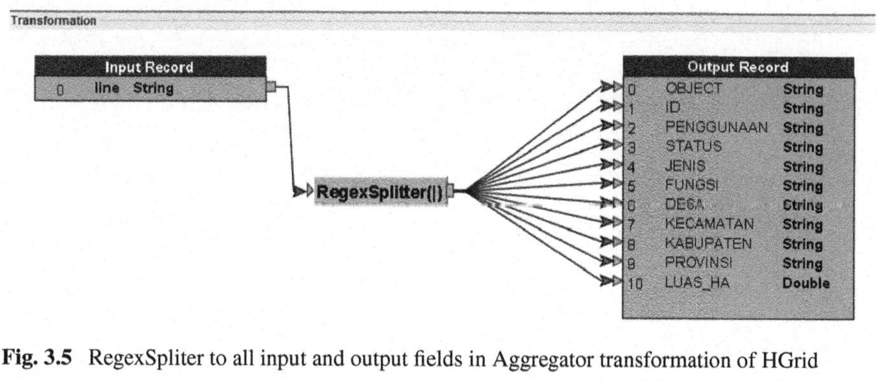

Fig. 3.5 RegexSpliter to all input and output fields in Aggregator transformation of HGrid

METER as the parameter unit (Fig. 3.6). The result of HDFS flat File saved and workflow can be processed. File_mr.jar will be created, and there is a class called workshop.ShpProcess in this. The part of the output will be shown in Fig. 3.7.

From 1st test of processing IG_Garam data in determining the distance of all points to a certain point (ID 625,745), it is known that there are two points whose distance is 0 (zero) because they overlap. To find the distance of the polygon, we used the centroid to find the distance between the center points (center to center), distance to distance (the outermost part of the polygon). The outermost part of the polygon to the surrounding polygons was used, and earlier, we used the unit m, so the distance in meters means less than 1 m (cm). If there is data that does not meet

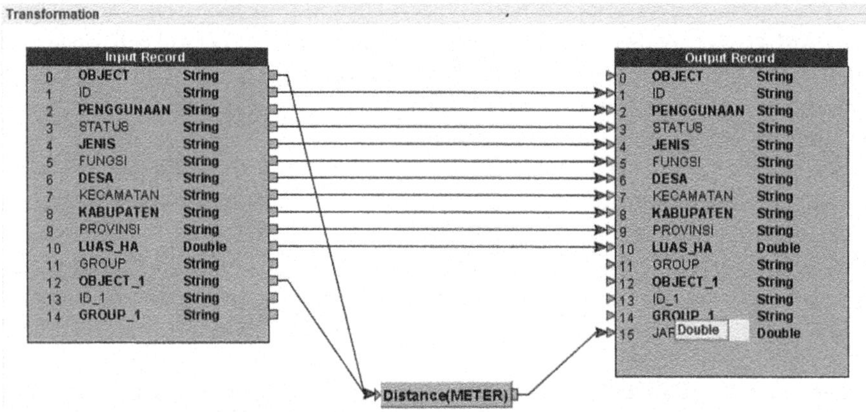

Fig. 3.6 Aggregator transformation of 1st case

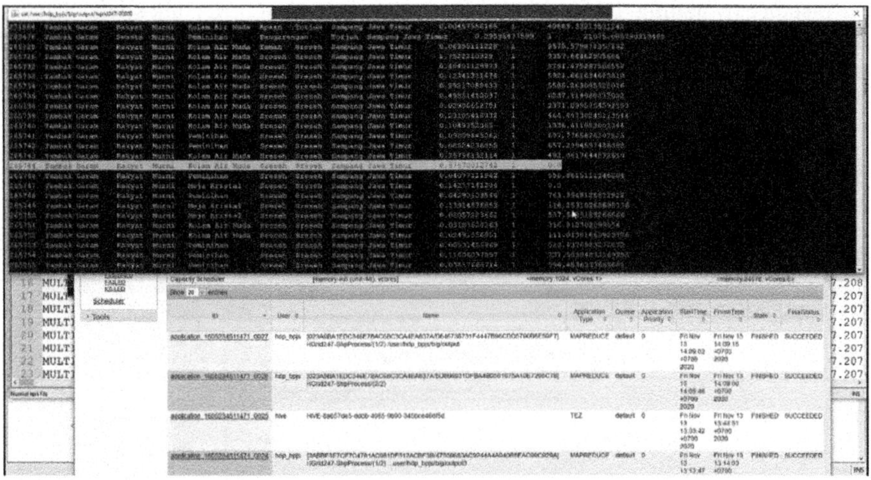

Fig. 3.7 Distance output of 1st case

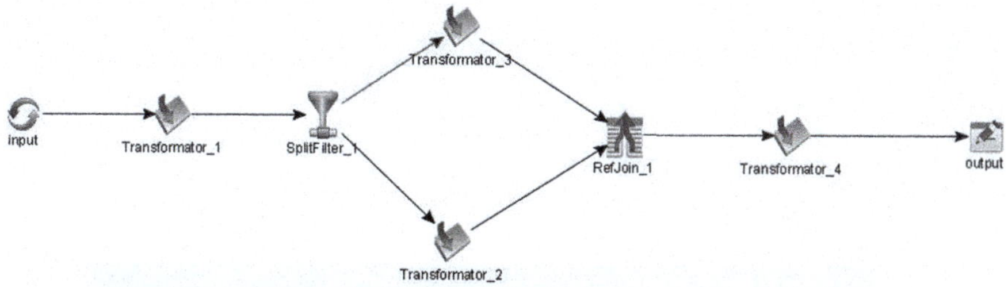

Fig. 3.8 Example of the failed flow process

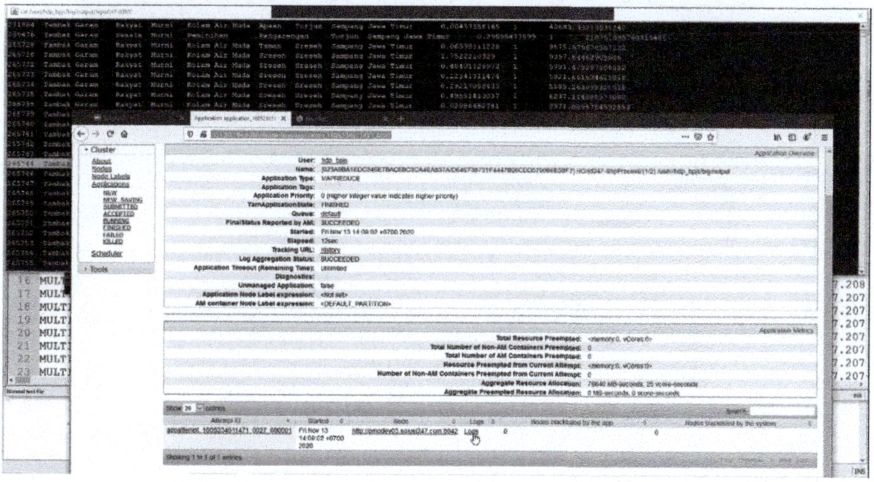

Fig. 3.9 Application report of 1st case

the requirements, then the flow process becomes red (Fig. 3.8), and if an error occurs (final status error or unsuccess), it can be checked from the log, depending on what happened, including whether a new library is needed (Fig. 3.9).

The second experiment using landslides data from Bogor Regency with the initial copy process of data to the main server to processes HDFS. Same with the previous process using salt mapping identification (IG_Garam), but using Union Function of two geometries must have the same dimension (Fig. 3.10), which always returns the result as a collection (Fig. 3.11).

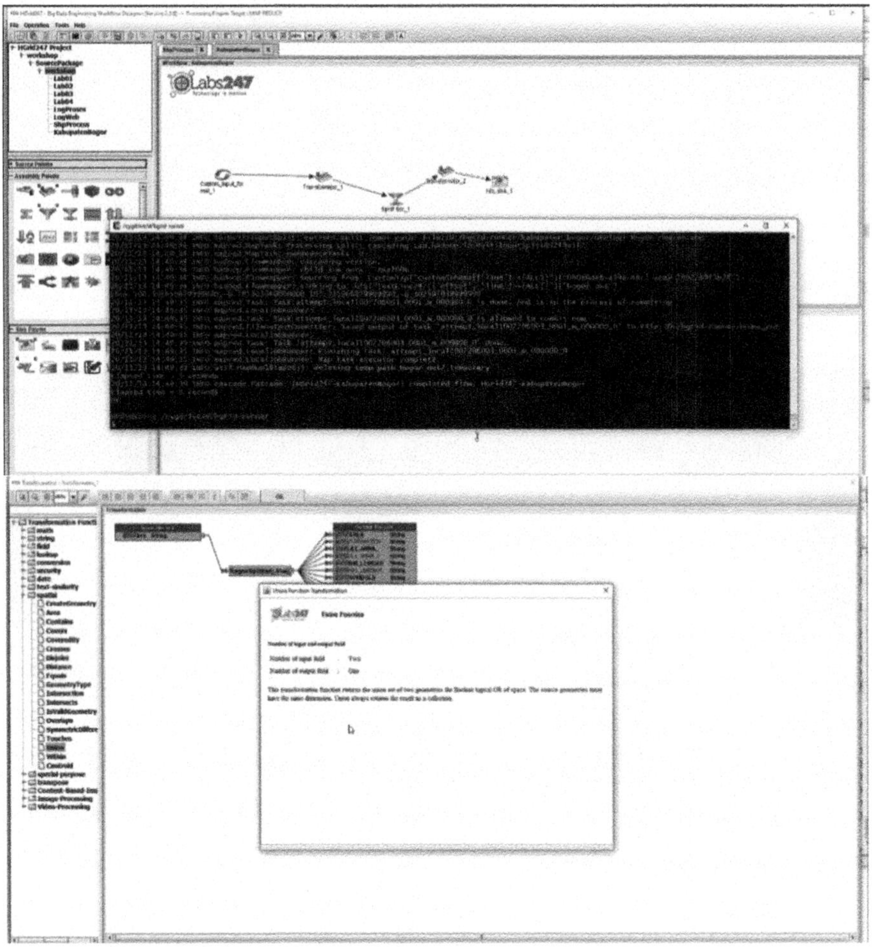

Fig. 3.10 An example of the union process of the second case

3.4 Discussion

Characteristics of data-driven in big data could be (1) data is centralized and organized to ensure data is recent, relevant, and data coming from across the organization, (2) data is accessible, (3) data governance policies are in place to safeguard the quality of critical master data objects like the product, customer and supplier, there should be transparent data governance process, and (4) analytics is integrated into tools to define analytics models easily. The problem begins with more business questions are required to answer, faster reports and Analytics are required, and more data is required [16].

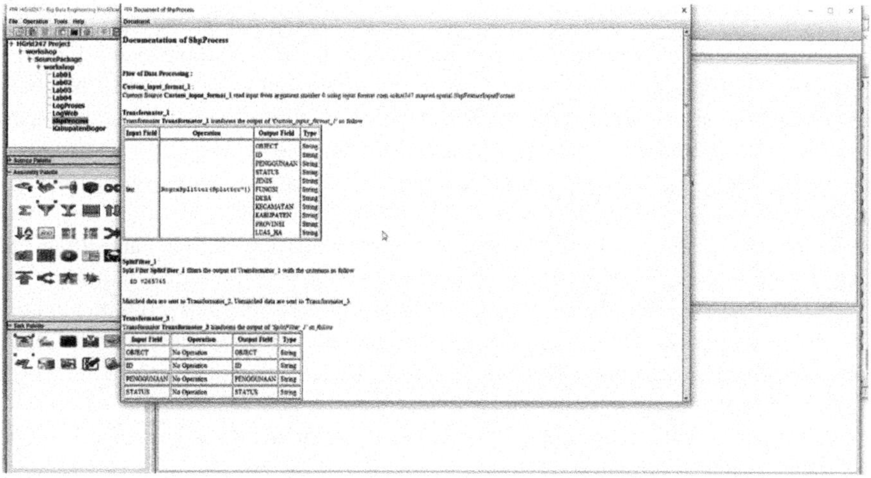

Fig. 3.11 An example of the final report of HGRID for union process of the second case

GIS and Big Data are hotly debated topics in recent years, especially in its collection, access, and storage because of massive growth volumes of 2D and 3D spatial data and Spatio-temporal data, along with an effective manner to map, analyze, and interpret the massive datasets. Real-time data are received as a stream and used to build up a dynamic map or generate statistical values by implementing sequential analysis cumulatively [17]. In analytic tools and computational techniques of big data, it relies on different research methodologies, tools, and computational models, i.e., meta-network modelling, network discovery and network analysis methodologies, dynamic network analysis and statistical prediction, agent-based simulation models, behavioral sequencing techniques, and genetic computation, collaborative and automated coding tools for unstructured text data [10, 18].

Data lake raised to help store a vast amount of data inexpensively, collect all data, fresh up expensive EDW resources, and explore data any time, easily scalable, and some processing is better done in Hadoop [7, 8]. We found that YAVA is user-friendly in managing geospatial big data, and its stack component includes HDFS, MapReduce, Yarn, Hbase, and Hive. However, YAVA does not provide a public online repository, so a local repository must be created. HBase is a type of "NoSQL" database; it is not an RDBMS. HBase is a distributed database, so technically speaking, HBase is more a "Data Store" than a "Database" because it lacks many features we find in an RDBMS. ClusterBy and DistributeBy are used mainly with the Transform/Map-Reduce Scripts in selecting the syntax. Besides, Hive uses the columns in Distribute By to distribute the rows among reducers. All rows with the same Distribute By columns will go to the same reducer. In our testing, the spatial menu in HGRID is helpful with limitations in visual-spatial view, and the output report is informative enough.

There is no specific rule for server node recommendation. However, some guidance can be used as follow (1) for master node (1 GB RAM for every 1 million blocks of HDFS) and Slave (1 core CPU with storage intensive is higher than 6 TB of storage, balance workload is typically 3–6 TB of storage, and compute-intensive are 1–3 TB of storage), (2) price consideration are 4 × 6 core CPU versus 3 × 8 core CPU, and (3) power consideration for electricity takes 30–50% of the total data centre cost (including cooling system). In balance workload, for more Disk need indexing, grouping, export-import data, data transformation, or cold data storage use case for more CPU we need machine learning processes, complex text mining and NLP, real-time, in-memory process. If the data growth per day is 4 TB with retention in two years period, for storage sizing requirement estimates data growth is 10% per month, with assumption using three replications, so the total for storage will be 4 TB × 3 replications × 12-month × 2 year reaches 288 TB. With the assumption of 28 TB per node for a raw capacity in balance workload, we estimated the number of nodes is 288 TB divided by 28 TB per node reaches 10.29 nodes or around 11 nodes. It recommends separating master and worker nodes to isolate task/application workloads, and slaves nodes are frequently decommissioned for maintenance.

3.5 Conclusion

Geospatial Big Data represents big data in the form of layers and spatial attributes [7]. There are countless numbers of distributed data. Technology enables stakeholders to collects, analyze, store, and disseminates geographic information for use by different entities to facilitate informed and smooth decision making. YAVA Hadoop and HGRID help manage geospatial big data. YAVA Hadoop leverages the power of the Apache Hadoop ecosystem and is designed to help accelerate Hadoop deployments and their ecosystem in Indonesia. Meanwhile, HGrid is an open-source software framework for storing data and running applications on a locally made group of commodity hardware or a big data engineering tool (ETL Tool) to process big data automatically. The spatial menu in HGRID, i.e., Create Geometry, Area, Contains, Covers, Covered by, Crosses, Disjoint, Distance, Equals, Geometry Type, Intersection, Intersects, Overlaps, Union, Within, and Centroid. The spatial menus of Hagrid247 are helpful. However, it is not a visual-spatial display, but rather a data engineering, case by case use of the tools according to the purpose, but the output report is informative enough.

Acknowledgements We thank our colleagues from the Center for Management and Dissemination of Geospatial Information of the Geospatial Information Agency of Indonesia and PT. Solusi 247, who provided related documents and insights and expert inputs that greatly assisted the research. However, they may not agree with all of the interpretations/conclusions of this paper. This work was supported by the Geospatial Information Agency of Indonesia (Badan Informasi Geospasial) thru DIPA 2021. The funders of the study had no role in data analysis and processing or writing of the report.

References

1. Jin, X., Wah, B.W., Cheng, X., Wang, Y.: Significance and challenges of big data research. Big Data Res. **2**(2), 59–64 (2015). https://doi.org/10.1016/j.bdr.2015.01.006
2. Lee, J., Kang, M.: Geospatial big data: challenges and opportunities. Big Data Res. **2**(2), 74–81 (2015). https://doi.org/10.1016/j.bdr.2015.01.003
3. Li, S., Dragicevic, S., Antón, F., Sester, M., Winter, S., Coltekin, A., Pettit, C., Jiang, B., Haworth, J., Stein, A., Cheng, T.: Geospatial big data handling theory and methods: a review and research challenges. ISPRS J. Photogramm. Remote Sens. (2015). https://doi.org/10.1016/j.isprsjprs.2015.10.012
4. Robinson, A.C., Demšar, U., Moore, A.B., Buckley, A., Jiang, B., Field, K., Kraak, M., Camboim, S.P., Sluter, C.R.: Geospatial big data and cartography: research challenges and opportunities for making maps that matter. Int. J. Cartogr. **3**(sup1), 32–60 (2017). https://doi.org/10.1080/23729333.2016.1278151
5. Silalahi, F.E.S., Prayitno, M., Gambetta, W., Amhar, F., Wijaya, M.N.Q.A., Rachma, T.R.N.: Identifikasi Pertumbuhan Data dan Penggunaan Application Performance Index (APDEX) Score dalam Penilaian Kinerja Aplikasi Penyaji Informasi Geospasial (Data Growth Identification and Application Performance Index (APDEX) Evaluation on the Performance of Geospatial Information Mapping Applications). Jurnal PEKOMMAS (Penelitian Komunikasi, Informatika Dan Media Massa) **6**(April), 95–104 (2021). https://202.89.117.136/index.php/pekommas/article/view/2060110/pdf
6. President of The Republic of Indonesia.: Presidential Regulation of The Republic of Indonesia Number 39 Year 2019 regarding One Data of Indonesia (2019)
7. Li, Z.: Geospatial big data handling with high performance computing: current approaches and future directions. In: Tang, W., Wang, S., (eds.) High Performance Computing for Geospatial Applications. Geotechnologies and the Environment, vol 23. Springer, Cham (2020). https://doi.org/10.1007/978-3-030-47998-5_4
8. Ravat, F., Zhao, Y.: Data lakes: trends and perspectives. In: International Conference on Database and Expert Systems Applications (DEXA 2019), Linz, Austria (2019). https://oatao.univ-toulouse.fr/25043/
9. Zhang, F., Zhou, Z., Xu, W.: Distributed storage and processing method for big data sensing information of machine operation condition. J. Softw. **9**(10), 2598–2606 (2014). https://doi.org/10.4304/jsw.9.10.2598-2606
10. Amhar, F., Tampubolon, W., Silalahi, F.E.S.: Combination of non hadoop-bigdata & relational DB (pp. 405–410). In: 41st Asian Conference on Remote Sensing (ACRS 2020) (2020)
11. Chen, Q., Wang, L., Shang, Z.: MRGIS: a mapreduce-enabled high performance workflow system for GIS. IEEE Fourth International Conference on eScience **2008**, 646–651 (2008). https://doi.org/10.1109/eScience.2008.169
12. Daneshyar, S., Razmjoo, M.: Large-scale data processing using mapreduce in cloud computing environment. Int. J. Web Serv. Comput. (IJWSC) **3**(4), 1–13 (2012)
13. Fan, W., Bifet, A.: Mining big data: current status and forecast to the future. SIGKDD Explor. Newsl. **14**(2), 1–5 (2013). https://doi.org/10.1145/2481244.2481246
14. Wasser, L., Chris Holdgraf, M.M.: About the Geotiff (.tif) raster file format: raster data in python. Earth Data Sci. Earth Lab (2020). https://www.earthdatascience.org/courses/use-data-open-source-python/intro-raster-data-python/fundamentals-raster-data/intro-to-the-geotiff-file-format/
15. NASA Earth Data.:. GeoTIFF | Earthdata (2021). https://earthdata.nasa.gov/esdis/eso/standards-and-references/geotiff
16. Gästbloggare.: 4 characteristics of data-driven organizations—And how to get started. Analytics (2016). https://www.ibm.com/blogs/think/se-sv/2016/04/15/4-characteristics-of-data-driven-organizations-and-how-to-get-started/

17. Rahimi, M.M., Hakimpour, F.: Towards a cloud based smart traffic management framework. Int. Arch. Photogramm. Remote Sens. Spat. Inf. Sci. **XLII-4**(W4) (SMPR and EOEC, Tehran, Iran) (2017)
18. Brunswicker, S., Bertino, E., Matei, S.: Big data for open digital innovation—A research roadmap. Big Data Res. **2**(2), 53–58 (2015). https://doi.org/10.1016/j.bdr.2015.01.008

Chapter 4
Next-Generation Shopping Cart Design Based on Design Methods for Consumer Behaviour Analysis

Yu-Ching Hung, Yann-Long Lee, An-Ting Kiat, Ann-Yee Wong, Mei-See Chiah, and Jhong-Yan Cai

Abstract A shopping cart is a very important aid to help customers to transport their goods in the market and mall. Sometimes there will be insufficient for a shopping cart in the supermarket. And the shopping carts are always scattered around everywhere inside or outside supermarkets. Based on this, researchers found out most of the people faced difficulty controlling the shopping cart and used a bad condition of the shopping cart before. Besides, customers have also faced problems of lack of information on promotions, inconvenient payment processes, and do not clear about the product location. Most people also faced problems during the payment process as they required to queue for a long time. According to the statements above, the problems that can ruin customers' shopping experiences are the bad condition of a shopping cart, insufficient shopping cart, current payment method with low efficiency, and lack of information about product location and promotion. In this research, the project target to aim for the design 2040. Here use several design methods, such as product specification analysis, market positioning analysis, Morphological analysis, and KJ (Kawakita Jiro) method to define the market objective and problem-solving. Designers are going to design a product which is a smart shopping trolley that can benefit human life in 2040. Besides, this shopping cart was designed able to enhance the customers' shopping experiences and improve the shopping process by solving all the problems faced by customers. In conclusion, researchers achieve the objectives stated to design a future product that able to enhance the shopping experience in 2040 which is a smart shopping cart by using the design tools and industrial design principles.

Keywords Product design · Design methods · Design project · Future design

Y.-C. Hung · A.-T. Kiat · A.-Y. Wong · M.-S. Chiah
Faculty of Mechanical and Manufacturing Engineering Technology (FTKMP), Universiti Teknikal Malaysia Melaka, Melaka, Malaysia

Y.-L. Lee (✉) · J.-Y. Cai
Dept. of Industrial Design, Chaoyang University of Technology, Taichung, Taiwan
e-mail: yannlee@cyut.edu.tw

M. A. Laribi et al. (eds.), *Advances in Automation, Mechanical and Design Engineering*, Mechanisms and Machine Science 121,
https://doi.org/10.1007/978-3-031-09909-0_4

4.1 Introduction

Grocery shopping constitutes an essential and routine type of consumer behavior [1]. It is mutual thinking that the shopping experience starts when a shopper enters a store or an e-commerce website and stops when this same shopper leaves with its purchase [2]. However, some situations and conditions had affected customers' shopping experiences.

In some of the shopping retail, the shopping carts provided are in bad condition and this will ruin the customers' shopping experiences. For some people to drag a trolley can be a tough job [3]. The user needs to apply extra forces to move a shopping cart with a broken wheel in the desired direction. Sometimes, the brake system of the shopping cart will also bring trouble for the user. A cart will tend to toll down even the slightest slope when an attending customer eases or release his hold on the cart.

A shopping cart is a very important aid to help customers to transport their goods but sometimes there will be insufficient for a shopping cart in the supermarket. Some selfish people would like to steal the shopping cart from the shopping mall without any payment or permission. It was also found that shopping trolleys are always scattered around everywhere inside or outside supermarkets [4] since the users did not return the shopping trolleys to the original place. This situation is not only troublesome for the customers, but also cause the owner has to find these missing trolleys or purchase the new trolleys.

In addition, customers have also faced a problem due to the current payment method in which the cashiers have to scan the price of the goods one by one. Among the difficulties faced by the customers, one difficulty is to follow the queue through the billing process [5]. This tedious process which involves the scanning of barcodes employed by a large number of supermarkets increases the time and results in queuing delay [6].

The next problem that affects the shopping experience is the lack of information about product location and promotion. Some customers will miss up on the information about the promotion especially during the seasonal sales period. Problems would arise when the size of the shopping center is relatively large and sometimes customers do not know where certain items are placed [7]. It is known to be an inconvenience and time-wasting for customers who are in a rush to search for desired products in a supermarket [8]. Based on the above, this research would like to integrate some basic design methods to design a shopping cart for 2040.

4.1.1 Product Define

A trolley is a device used for transporting or carrying the load. There is various type of trolley for different applications. The trolley that used in the shopping mall or supermarket, it called a shopping trolley or shopping cart. It is different between a shopping trolley and a load trolley. To put it simply, the shopping trolley can be defined as the basket on the wheels. The shopping trolley consists of a basket or a bag to hold the object whereas the load trolley consists of a flat platform to carry heavy duty.

Before the invention of a shopping trolley, people used baskets the hold more things but they didn't hold much due to the weight of the things. The owner of the Humpty Dumpty Supermarket in Oklahoma, Sylvan Golden was searching for a solution to help the customers move more groceries. In 1937, he came up with a solution and developed a shopping cart to help the customers carry more groceries.

A shopping trolley is a basic tool for shopping in a physical store such as a shopping mall or supermarket. It is supplied by the owner of a shop so that the customers can use it to transport groceries as they move around the supermarket. Most of the shopping trolley is attached with a handled cane basket to carry objects. The wheels are attached at the bottom of the body of the shopping cart to ease the movement. The basket is usually made of metal and the handle of the shopping trolley is made of plastic. The shopping trolleys are usually cost between 75 and 150 dollars each and the price of a shopping trolley will change according to its material, size, brand, and features.

4.1.2 Problem Statement

There are problems occurred during the shopping process and these make people prefer online shopping more than shopping in the physical retail store. Firstly, a shopping cart with bad conditions leads to inconvenience to the user as the user needs to exert more force to control its movement. Besides, some selfish users who simply leave the shopping cart anywhere or bring it back home caused the insufficient number of the shopping cart. This will not only ruin the customers' shopping experiences but also troublesome the owner of the shopping cart. On the other hand, the current payment method which is slow and low efficiency caused the customers to waste their time queuing and searching for money. This problem also contributes to the payment counter area being jam-packed with people. Last but not least, the customers also faced the problem of a lack of information about the location of products and promotions.

In short, the problem that will affect the shopping experience are: (1) the shopping carts were not be integrated by technology and bad condition shopping cart [4, 8]. (2) Users' habits and behavior were ignored [1, 2]. (3) The design of the shopping cart does not integrate the progress of the times and the needs of users. The aims and objectives are: (1) To design a product that may bring benefits to humans in 2040. (2) To design a future shopping cart by integrated the design methods. (3) To design a product that can enhance the customer shopping experience and improve the shopping process.

In this project, researchers are going to design a product which is a smart shopping trolley that can benefit human life in 2040. Besides, this shopping cart is designed able to enhance the customers' shopping experiences and improve the shopping process by solving all the problems faced by customers.

4.2 Methods and Analysis

To start a product design, designers must need to familiar with the product function, market target, and user's needs. Based on it, this research will through the existing product analysis, market segmentation, and user intention survey to get the user needs. To use creative design methods like KJ (Kawakita Jiro) method in the function and structure design to generate the idea concepts. And use Morphological analysis to integrate the functions and structures in design shape.

4.2.1 Specifications Analysis

There has a lot of design in the market and through marketing research, designers can learn their design concepts and knowledge. A systematic analysis of existing or similar products should be made before designing and manufacturing new products. The results of the analysis will help us to design a new product or improve an existing product. They can also lead to the development of scientific concepts and principles.

According to existing products, there are 20 trolleys collected from different sources which are from the catalog, online research, and book. All of these trolleys collected have different characteristics and features. These trolleys also have different designs and made of different materials such as steel, plastic, and fabric.

After 20 trolleys are completely collected, teams go further on the specification analysis. The specification analysis data can lead to the development of scientific concepts and principles of new product design. There are 20 product specifications that have been analyzed. Most of the trolleys are made of steel or consist of steel components because steel is an alloy of iron and carbon which has high tensile strength, low cost, and good durability. Besides, aluminum and plastic are common materials to manufacture a trolley because both materials are low cost, lightweight, and good formability. In addition, the trolley also has a different size but the trolley in bigger size is more suitable for an adult since the bigger size trolley has more storage capacity although it is heavy and difficult to control.

4.2.2 Consumer Intention Analysis and Market Positioning

A good product, and must be able to move users, and a good design, must understand the user needs. To get the user's need, here conduct a user needs questionnaire survey. Section A of this questionnaire is used to collect background information. Section B is related to the opinion of respondents about shopping carts and shopping experiences. Section C is in the Likert scale form. It consists of 12 questions and these questions are related to the customer requirements of the smart shopping trolley.

The result of the data enables us to know the respondents' opinions about the specification for designing a smart trolley structure so that we can improve the design and appearance by focusing on the important criteria. According to the survey, we can clearly understand the ergonomics, cleanliness, safety, easy to use, convenience, multipurpose, and storage capacity are the main criteria of the shopping cart that customers concerned about since these criteria obtained the highest mark. Besides, the criteria with 4 marks such as appearance and security also will be considered. This shows the current version of the shopping trolley still has many security crises that we have to improve. Besides, we must consider all of these points in our product design planning to satisfy the customers' needs which can produce the best and perfect product.

A product position is the way a company's brand fits into targeted market segments relative to competitors. Through the product position clear, it will assist companies to established differentiated product benefits to stand out from competitors.

Here use 20 existing unique products collected to represent the market situation. And use 2 positioning maps to define the market situation. Positioning Map 1 is Storage Capacity VS Multipurpose. Positioning Map 2 is Outstanding VS Appearance. After evaluating the products and marks given, will generate 2 position maps. And base the results of the user need the survey to select the target market.

4.2.3 The KJ Method

The KJ method is a technique used to gather and prioritize ideas. In Fig. 4.1, it had categorized our idea into 5 groups which are the condition of the trolley, security on the trolley, price checker, location checker, and payment method. Based on the ideas to develop in these 5 categories, designers had figured out the main design concept of the shopping cart is to support price calculation and payment consists of a security system, good quality, and provide price checker and location navigator. This can help us to focus on our design concepts for the shopping cart.

4.2.4 Morphological Analysis

Invented by Professor Zwicky of the California Institute of Technology in the 1940s, it was first used in the field of mechanical engineering. This method lists all the independent elements of a project and lists its variable parameters for each independent element to make an analysis table. Finally, all the combinations will be studied in detail.

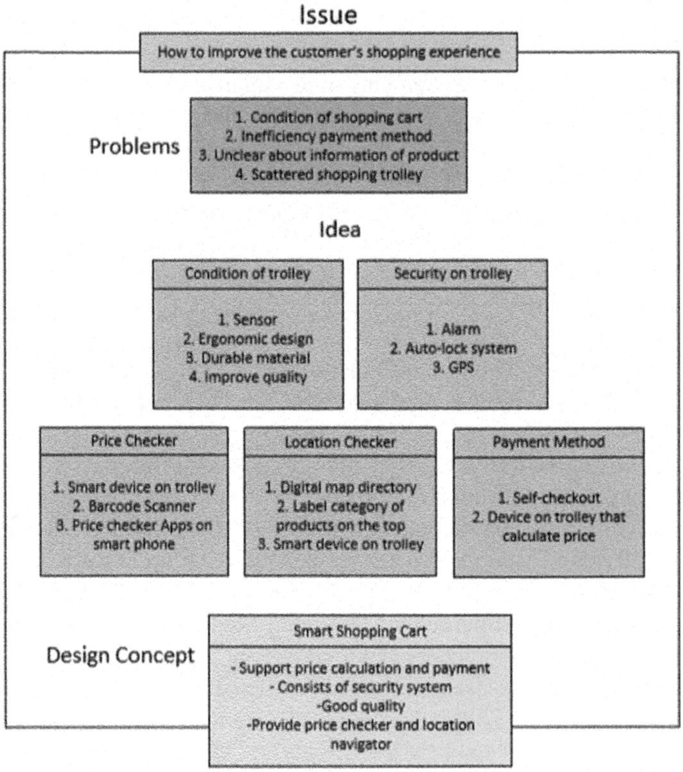

Fig. 4.1 KJ method

The morphological analysis is a table used to analyze the function, part, and design of a product. The details are presented in visual form to help the user to generate concept ideas. In this morphological chart, designers categorized the shopping cart into 5 important parts which are basket, handle, wheel, chassis material, and type of smart device. Each of the criteria comes along with 4 options. These options are based on some factors which are ergonomic, simple, and outstanding. The combination of the options can be used as a reference for the design concept. From Fig. 4.2.

4.3 Projects Design

Based on the questionnaire analysis, researchers found out most of the people faced difficulty controlling the shopping cart and used a bad condition of the shopping cart before. Besides, they have also faced problems of lack of information on promotions, inconvenient payment processes, and do not clear about the product location. Most

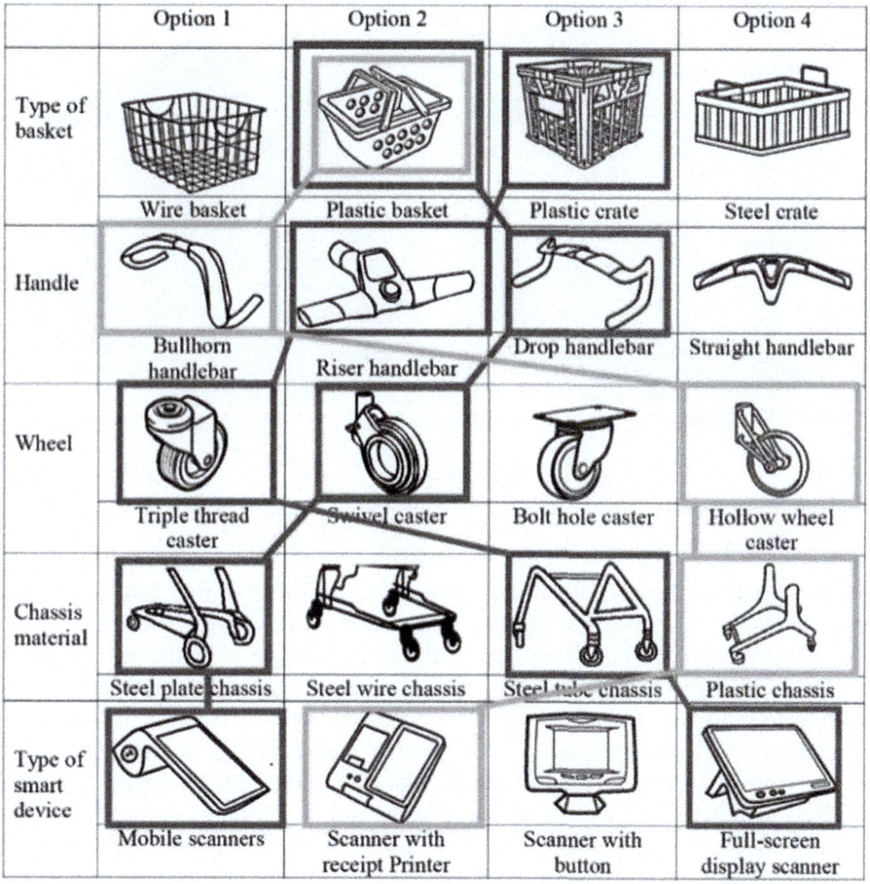

Fig. 4.2 Morphological chart

people also faced problems during the payment process as they required to queue for a long time.

4.3.1 Concept Design

In this project, a total of 7 shopping carts were designed. Due to the length of the article, only 2 of them were selected for explanation.

Concept 6: This concept design is based on the combination of the rickshaw and the tablet. All the actions of the trolley can be controlled by the tablet such as member login, payment, information access, and navigation. To ensure the freshness of the groceries, this design provides a freezer at the front part of the trolley with a simple

Fig. 4.3 The 3D model drawing of concept 6 (Left), the 3D model drawing of concept 7 (Right)

push and pull lid. Besides, the trolley provides space to install the customer's bag to ease the packing process. Lastly, the trolley has a stabilizer with a customized logo that helps to stabilize the movement of the trolley. The handle and the size of the trolley are designed ergonomically to protect the customers from any accident and injuries (See Fig. 4.3).

Concept 7: The design concept of this sketch is based on the backhoe. The chassis of the trolley is a design based on the backhoe's arm this chassis is used to support the basket so that it can withstand heavy load steadily. Furthermore, the back wheel of the trolley is larger than the front wheel this is because the bigger back wheel can avoid slipping. However, a smaller radius of the front wheel can help to turn a sharp corner easily. Moreover, this design also has an ergonomic design handle in which the shape of the handle is based on a bicycle handle. Then, the basket of this trolley is designed in aerodynamics shapes, this design helps in the movement of the trolley by reducing air friction. Besides, the scanner device is attached between the handle and basket with a degree that able the user see the price easily. Last but not least, the wheels will break automatically when it is out of the designated area (See Figs. 4.3 and 4.4).

4.4 Conclusion

In this project design, designers achieve the objectives stated to design a future product that able to enhance the shopping experience in 2040 which is a smart shopping cart by using the design tools and industrial design principles. Through this project, designers gain a lot of knowledge about industrial design. Firstly, the

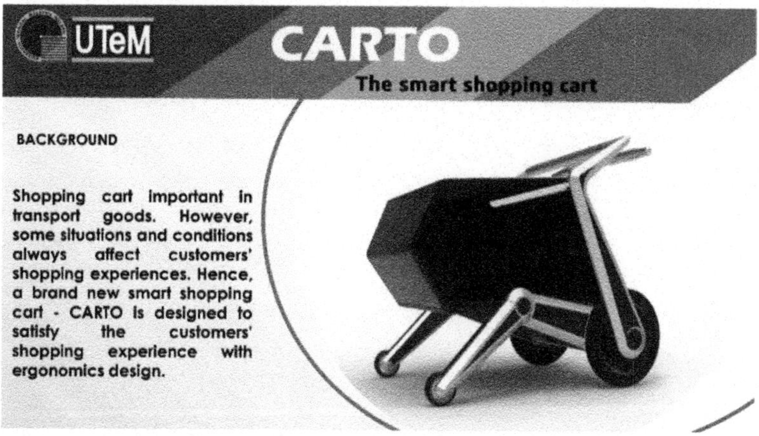

Fig. 4.4 Exhibition design of concept 7

skill of sketch in industrial design subjects is important. Clearly understanding the concept of perspective view and makes a lot of practices will enhance the industrial design sketch skill and sketch quality. Besides, the design tools such as brainstorming, questionnaire analysis, morphology analysis, and others play an important role in the design process. These several tools assisted us in the design process by coming out with more ideas and improve the design.

Acknowledgements Thanks for the funding support from the Ministry of Science and Technology, Project ID: MOST 110-2410-H-324-006.

References

1. Whan Park, C., Iyer, E.S., Smith, D.C.: The effects of situational factors on in-store grocery shopping behavior. J. Consum. Res. **15**(4), 422–433 (1989)
2. Hussain, M.M., Siddiqui, D.A.: Influence of shopping experience on shopping/convenience store selection. Int. J. Bus. Adm. 10(3), 74–89 (2019)
3. Luk, S.T.K., Sharma, P., Chen, I.: Shopping motivation as a moderator in the retail service evaluation. J. Serv. Mark. 27(1) (2013)
4. Ng, Y.L., Lim, C.S., Danapalasingam, K.A., Tan, M.L.P.: Automatic human guided shopping trolley with smart shopping system. Jurnal Teknologi 73(3) (2015)
5. Manikandan, T., Mohammed Aejaz, M.A., Nithin Krishna, N.M., Mohan Kumar, A.P., Manigandan R.: RFID based advanced shopping trolley for super market. J. Chem. Pharm. Sci. JCHPS (Special Issue 8), 225–230 (2017)
6. Slack, N., Singh, G., Sharma, S.: Impact of perceived value on the satisfaction of supermarket customers: developing country perspective. Int. J. Retail. Distrib. Manag. 48(11) (2020)
7. Ariffin, S.K., Mohammad Shah, K.A., Ismail, I.: Muslim consumers' attitudes toward the advertisement of non-certified coffee shops. Jurnal Pengurusan 48 (2016)
8. Sharma, P., Ueno, A., Kingshott, R.P.J.: Self-service technology in supermarkets—Do frontline staff still matter? J. Retail. Consum. Serv. 59 (2020)

Chapter 5
Automatic Modulation Classification Based on Machine Learning

Yilin Sun and Eddie Ball

Abstract Automatic Modulation Classification (AMC) is a rapidly evolving technology, which can be employed in software defined radio structures, such as for military communication. Machine Learning can provide novel and efficient technology for modulation classification, especially for systems working in low Signal to Noise Ratio (SNR). For this work, a dynamic modulation classification system without phase lock is trialed. The signals are captured with different SNR and duration. Traditional Machine Learning based on the mathematical features is compared with Deep Learning based on the constellations. Based on these two methods, a hybrid model is provided. This model involved the novel Deep Learning at first and the feature classification as a supplement, which achieves good performance at low SNR area.

Keywords Automatic modulation classification · Machine learning

5.1 Introduction

As digital communication systems develop, improving radio spectrum usage efficiency becomes vital. To respond to this requirement, Dynamic Spectrum Access (DSA) is a good starting point, with spectrum sensing and signal classification. In this case, the modulation classification performs a significant role and can be widely employed in a variety of applications, such as software defined radio system and radar communication in the military. There is a high demand of radio frequency (RF) bands. In the crowded spectrum situation, the AMC technology can respond to the requirements, optimizing signal demodulation, information extraction and interference detection without limitation from various complex emitters [1, 2].

This work concentrates on modulation classification provided by different methods in the machine learning area. With the improvement of the Artificial Intelligence (AI) technology, many areas have achieved new progress by this novel area. For

Y. Sun (✉) · E. Ball
The University of Sheffield, Sheffiels, United Kingdom
e-mail: Ysun57@sheffield.ac.uk

M. A. Laribi et al. (eds.), *Advances in Automation, Mechanical and Design Engineering*, Mechanisms and Machine Science 121,
https://doi.org/10.1007/978-3-031-09909-0_5

the traditional statistical methods of machine learning, classification of the modulation types by the statistical features of the signals is common. The most suitable model will be selected after the experiments of the calculation and comparison among SVM, K-Nearest-Neighbors (KNN), Decision Tree, and so on. After that, the Deep Learning working as a subproject of machine learning reached a new step, which extracts the essence from the biological information processing system. The advantages are also obviously shown in audio recognition, image classification and semantic segmentation. In this paper, the theory of the image classification will be utilized, Convolutional Neural Network (CNN) of deep learning is introduced and tested. Based on this novel method, the feature classification based on machine learning works as a supplement to improve accuracy. This new hybrid model detects the modulated signals from −10 dB SNR to 20 dB SNR.

In the previous research, Maximum-likelihood decision theory is used as a critical method. PSK and QAM signals are distinguished with accuracy of 90% above 9 dB SNR [3]. For the pattern algorithms, the features of signals are involved in the models, where high order cumulants play a critical role in the AMC algorithm. After extracting the efficient features, the Support Vector Machines (SVM) is applied in the recognition process. The accuracy can reach to 96% around 10 dB SNR for 200 samples [4]. But the probability of correct classification is between 50 and 70% at around 0 dB SNR, which still needs to be improved. Compared to the previous classifiers, the binary hierarchical polynomial classifiers are also proposed with the probability of correct classification of 65% [5]. There is still the challenge to optimize the AMC system, especially at low SNR.

This work will introduce the machine learning and deep learning methods. Images are captured by the dynamic system and used as training data. In this paper, the CNN model with the best accuracy and the high order cumulants will be involved in the hybrid system. The results at low SNR will also be analyzed in the following works.

5.2 Problem Presentation

In this section, the model of the signal, the statistical features and the constellations are introduced. The statistical features are used for the machine learning models and the hybrid model. The constellations of the modulated signals are utilized for the deep learning models and the hybrid model.

5.2.1 *Signal Model*

The received signal in baseband is defined with r(t) and it is given by

$$r(t) = s(t) + n(t) \tag{5.1}$$

$s(t)$ is the original signal which transmit through the additive white Gaussian noise (AWGN) channel.

With the demand of features calculation and analysis, the raw data should be represented by in-phase and quadrature components

$$a[i] = a_I[i] + j * a_Q[i] \tag{5.2}$$

In this case, signals simulated in software are composed of real part and imaginary part representing the characters of their constellation diagrams. All the statistical features and constellations are calculated and captured by this model. Furthermore, the data captured by hardware is also read as I/Q data. Building the model in this way can simulate the signals expected in reality.

Four kinds of modulated signals, BPSK, QPSK, 8PSK and QAM16 are applied in this work. There are 100 symbols utilized, sampled by the rate of 50 samples per symbol.

SNR is also an important index to describe the noise and imitate the signals in the real world. And it is defined by

$$\mathrm{SNR} = \frac{\mathrm{powerofsignal}}{\mathrm{varianceofnoise}} = 20\log_{10}\left\{\frac{\sqrt{\frac{1}{N}\sum |a[i]|^2}}{\mathrm{std}(|\mathrm{noise}|)}\right\} \tag{5.3}$$

In this work, signals are considered from −10 dB to 20 dB SNR. According to the previous works, focus was on the classification above 10 dB SNR [6]. The SNR value less than 0 dB should also be tested, to improve the performance of distinguishing modulation types. We have developed the models from −10 dB to 20 dB SNR to provide a comprehensive dataset.

5.2.2 *Statistical Features for Machine Learning and Hybrid Model*

The traditional statistical methods of machine learning are proposed in the previous research to classify the digital modulations [7]. The features proposed in the literature for artificial intelligence technologies are now considered. The statistical models can obtain the results by capturing the statistical features among the four modulation types from −10 dB to 20 dB SNR.

The useful features are as follows.

Signal power ratio of in-phase and quadrature part β

$$\beta = \frac{\sum_n a_Q^2[i]}{\sum_n a_I^2[i]} \tag{5.4}$$

Standard deviation of the direct instantaneous phase σ_{dp}

$$\sigma_{dp} = \sqrt{\frac{1}{N}\left(\sum\nolimits_{a_n[i]} \varphi_{NL}^2[i]\right) - \left(\frac{1}{N}\sum\nolimits_{a_n[i]} \varphi_{NL}[i]\right)^2} \quad (5.5)$$

where N is the number of the samples, $\varphi_{\mathrm{NL}}[\mathrm{i}]$ is the instantaneous phase, which is defined by $\varphi_{\mathrm{NL}}[\mathrm{i}] = \tan^{-1}\frac{\mathrm{a_Q[i]}}{\mathrm{a_I[i]}}$.

Standard deviation of the signal instantaneous normalized amplitude σ_{aa}

$$\sigma_{aa} = \sqrt{\frac{1}{N}\left(\sum\nolimits_{i=1}^{N} a_{cn}^2[i]\right) - \left(\frac{1}{N}\sum\nolimits_{i=1}^{N} |a_{cn}[i]|\right)^2} \quad (5.6)$$

$$a_{cn}[i] = \frac{a[i]}{E(a[i])} - 1 \quad (5.7)$$

Standard deviation of the signal normalized amplitude of signal σ_v

$$\sigma_v = \sqrt{\frac{1}{N}\left(\sum\nolimits_{i=1}^{N} a_v^2[i]\right) - \left(\frac{1}{N}\sum\nolimits_{i=1}^{N} |a_v[i]|\right)^2} \quad (5.8)$$

$$a_v[i] = \sqrt{\frac{a[i]}{var(a[i])}} - 1 \quad (5.9)$$

Standard deviation can describe the statistical dispersion, which the above three kinds of standard deviation results are utilized for describing the signals in the main characteristics of phase and amplitude.

Mixed order moment v_{20}

$$v_{20} = \frac{M_{42}(y)}{M_{21}^2(y)} = \frac{E(|a[i]|^4)}{E(|a[i]|^2)} \quad (5.10)$$

Mixed order moment is defined by the fourth order moment M_{42} and second order moment M_{21} of the signal. This feature employs the Joint Power Estimation and Modulation Classification (JPEMC) algorithm, which is associated with the power of signal and noise. This feature can reflect the power in another way [7].

Mean value of the signal samples X

$$X = \frac{1}{N}\sum\nolimits_{n=1}^{N} |a[i]| \quad (5.11)$$

Normalized square root of signal X_2

$$X_2 = \frac{\sqrt{\sum_{i=1}^{N} |a[i]|}}{N} \tag{5.12}$$

This feature can describe the amplitude scale of the signals.

Maximum value of power spectral density (PSD) γ_{max}

$$\gamma_{max} = \frac{1}{N} max|DFT(a_{cn}[i])|^2 \tag{5.13}$$

This feature describes the situation of signals in the frequency domain, by using the Discrete Fourier Transform (DFT).

Cumulants

$$C_{20} = E[a^2[n]] \tag{5.14}$$

$$\mathrm{C_{21} = E[|a[n]|^2]} \tag{5.15}$$

$$\mathrm{C_{40} = M_{41} - 3M_{20}^2} \tag{5.16}$$

$$\mathrm{C_{41} = M_{41} - 3M_{20}M_{21}} \tag{5.17}$$

$$\mathrm{C_{42} = M_{42} - |M_{20}|^2 - 2M_{21}^2} \tag{5.18}$$

$$\mathrm{C_{63} = M_{63} - 6M_{20}M_{40} - 9M_{42}M_{21} + 18M_{20}^2M_{21} + 12M_{21}^3} \tag{5.19}$$

$$\mathrm{C_{80} = M_{80} - 35M_{40}^2 - 28M_{60}M_{20} + 420M_{40} - 630M_{20}^4} \tag{5.20}$$

$M_{p+q,p}$ could be defined by $E\left[a[n]^p a[n]^{*q}\right]$. Cumulants and moments are widely used in variety of classification, especially for the high-order cumulants, which can avoid the influence from the AWGN [9, 10]. Cumulants are proposed to represent the moments in an alternative way and the moments can measure functions quantitatively. When the higher order cumulants of the received signals are calculated, the part of the Gaussian noise is zero and the signals are independent [11].

Kurtosis K

$$\mathrm{K} = \left| \frac{\mathrm{E[(a - E[a])^4]}}{\mathrm{E[(a - E[a])^2]^2}} \right| \tag{5.21}$$

Kurtosis is a feature which can describe the steepness or flatness of the distribution of signals.

Skewness

$$S = \left| \frac{E[(a - E[a])^3]}{E[(a - E[a])^2]^{3/2}} \right| \tag{5.22}$$

Skewness can describe the position of the tapering side of the distribution, which is also the third central moment. Both the abilities of kurtosis and skewness are to confirm the shape of the signals [6].

Ratio of peak-to-rms, PR

$$PR = \frac{\max|a|^2}{\frac{1}{N}\sum_{i=1}^{N}(a[i])^2} \tag{5.23}$$

Ratio of peak-to-average, PA

$$PA = \frac{\max|a|}{\frac{1}{N}\sum_{i=1}^{N}a[i]} \tag{5.24}$$

These two features aim to describe the shape of different signals in detail, which is not clear to observe.

Several of these technique 76 s are utilized in the traditional machine learning. To develop the hybrid model, the feature variables as supplements should be stable and clear to classify the modulation types.

5.2.3 Constellations of Modulation Format for Deep Learning and Hybrid Model

For the Deep Learning model and the first part of the hybrid model, the main principle is image classification. To be applied by the Deep Learning, the data of modulated signals are transferred to the constellation images. The Fig. 5.1 shows the constellations as examples of the four modulations at 10 dB SNR.

We assume that the systems providing the signals have phase and symbol lock. The SNR effects and constellations are directly observed from Fig. 5.1, which indicates the CNN should give a high accuracy of classification at 10 dB SNR. These constellations are captured in JPG format and stored with the labels of the modulation names. To get the efficient constellation images for the Deep Learning, the axes are limited from −4 to 4 of both directions. In this way, the sampled dots can stay in the main part of the images and the constellation images can have the same specification. There are 1200 images captured to feed the Deep Learning models. All the constellation images are used as dataset. 70% of the images are utilized for training, and the rest of images are for the validation process. The models will adjust the parameters to achieve the best performance based on the results of each experiment.

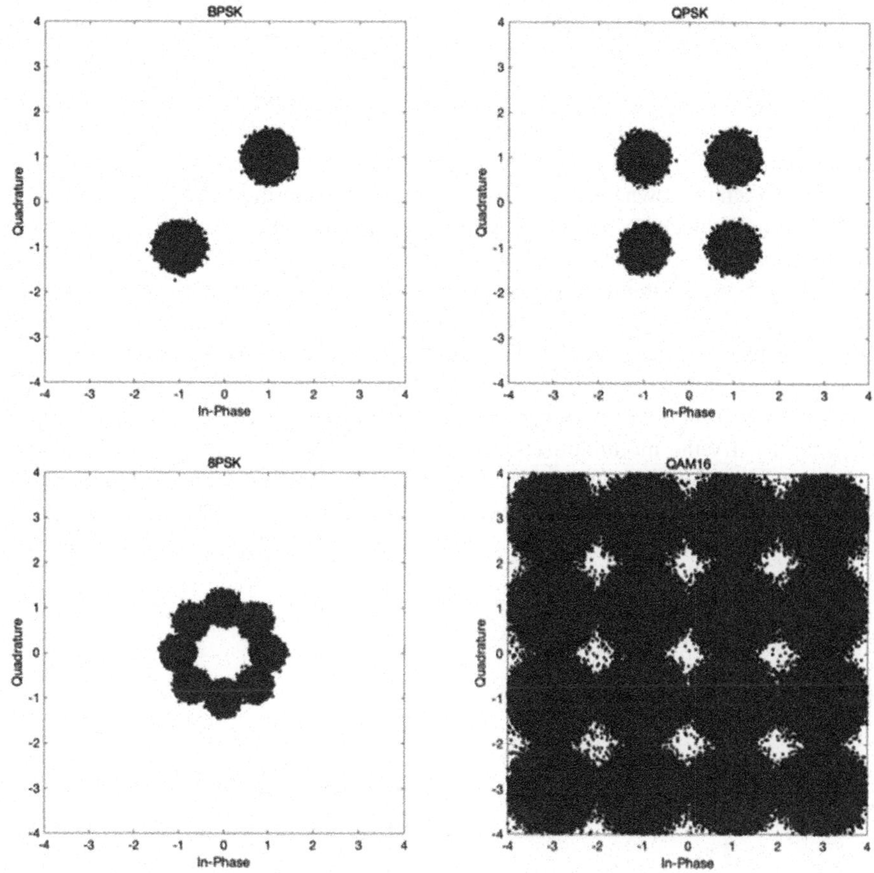

Fig. 5.1 Constellation of modulated signals with noise at 10 dB SNR

5.3 Classification Method

In this section, the classification methods are introduced. Machine Learning models are based on the statistical features from the modulated signals. The Deep Learning models achieve the classification according to the constellation images. Both of these two methods have limitations of the classification, the hybrid model is developed by combining them both.

5.3.1 Machine Learning

For the Machine Learning methods, the statistical features needed to be analyzed and selected to work in the model. Based on the formula of features from Sect. 5.2.2, the features from four kinds of modulation types are calculated.

Figure 5.2 has shown the statistical features of the four modulated signals at 10 dB SNR. To show the characters of all the features in one graph, some of the features are displayed in logarithmic form, such as σ_{aa}, X_2, γ_{max}, which is a clear way to apply the data with magnitude difference. According to the following graph, it is obvious that β, σ_{dp}, σ_v and X_2 cannot help to classify the modulations (BPSK, QPSK, 8PSK and QAM16). These four features remain constant around the fixed values.

The classification results from SVM, KNN and Decision Tree are compared. SVM is a supervised model that makes the classification between two groups. It can be used to classify the modulation types step by step. KNN is a supervised and non-parametric method, which need to set the value of k at first, the input consists of the k closest training examples in the feature space, the output is the class membership. The decision tree consists of a structure with different nodes which works as a condition

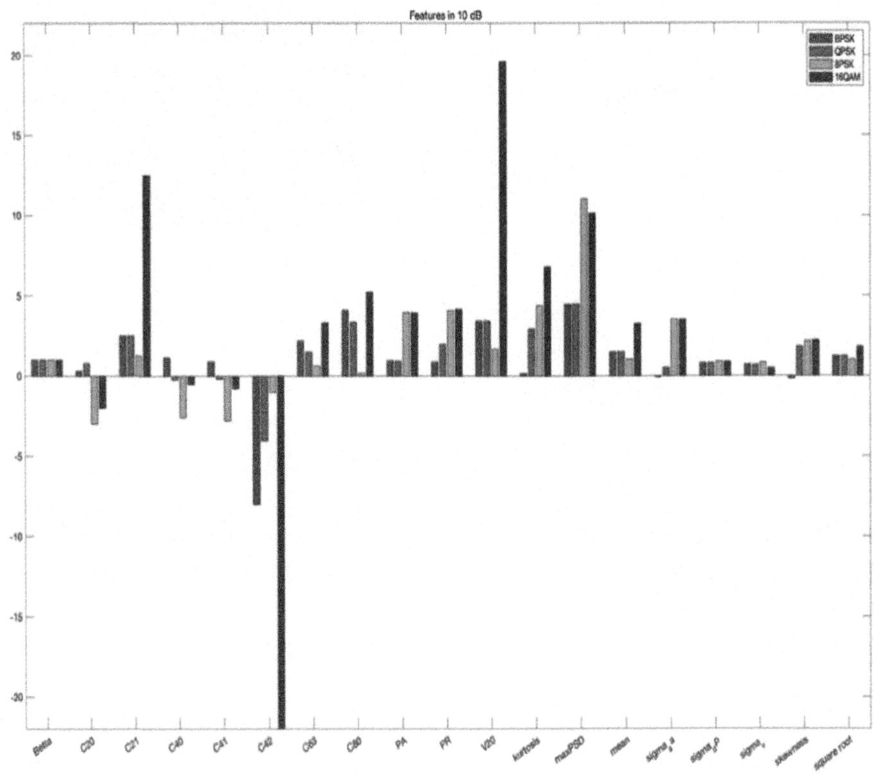

Fig. 5.2 Features of four kinds of modulated signals at 10 dB SNR

and check if the input data are satisfied. All of them are classic statistical methods to make decisions between the four types of modulation using features. But the calculation results from the signals fluctuate when the SNR varies, making decisions difficult.

5.3.2 *Deep Learning*

In this session, four CNN models are introduced. A simple CNN developed from the Iris case with several layers, SqueezeNet model, GoogleNet model and Inception-v3 model [12]. The last three models are pretrained networks, which are used for Transfer Learning. As Fig. 5.3 shown, an example of a simple CNN structure can consist of two convolutional layers and two dense fully connected layers. The characters of the signals can be captured by the two convolutional layers and the feature map can be produced by these two layers. After that the features can be learned by the two sequential layers. At last, the results of the classification are outputted.

CNN. This model is created by developing from the Iris recognition case. There are fifteen layers in total. Three convolutional layers are followed by the batch normalization layers and ReLu layers. Two max pooling layers are set between the other three blocks. The convolutional layers can capture the features of the images by the filters. The batch normalization layers can normalize the input channel. The ReLu layers calculate the threshold to the elements.

Transfer Learning Models. The SqueezeNet has 68 layers and its coefficients of the network were already pretrained to classify 1000 categories. The input image size is 227-by-227. The GoogleNet has 144 layers. It was pretrained to classify the images into 1000 categories and 365 places. The input image size is 224-by-224. The Inception-v3 has 315 layers and it was trained by more than a million images to classify into 1000 categories. The input size is 299-by-299. Images need to be resized before feeding to models. The last learnable layers are modified, to provide only 4 outputs, corresponding to each of the modulation types.

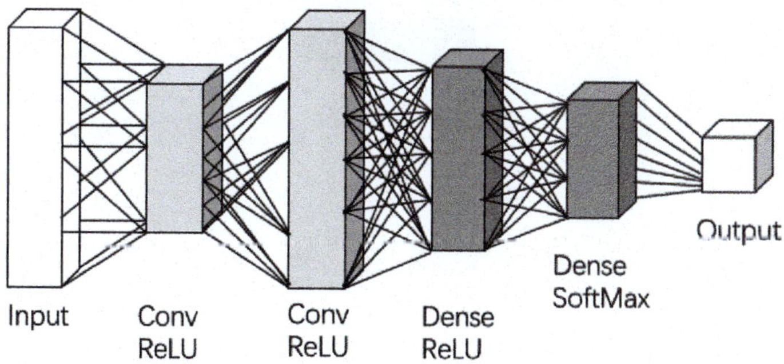

Fig. 5.3 Structure of CNN

In these experiments, 70% of images are used as training and the rest are used as validation. The augmented image datastore is applied to rotate the images from −90° to 90° (representing constellation phase errors) and rescale images from 1 to 1.5 randomly, which can help improve the amount of training data and avoid overfitting.

5.3.3 *Hybrid Model*

According to previous research, the hybrid method is provided. The deep learning is utilized at first, the statistical features of the signals are applied to the second process for the classification.

Figure 5.4 illustrates the probability of the predictions when the Inception-v3 model detects the signals at −3 dB SNR. It is obvious the system cannot classify the QAM16 from the BPSK and QPSK, so the statistical features must be involved to detect the proper detected signals.

The proposed structure of the hybrid model is shown in Fig. 5.5. The structure of deep learning is the model that gets the best classification results from the four CNN. We can see the results at low SNR are not accurate enough. Based on the traditional Machine learning method, the features which can obviously help distinguish the modulation types are chosen to help classify as continue. Feature classification requires the statistical features from the modulated signals.

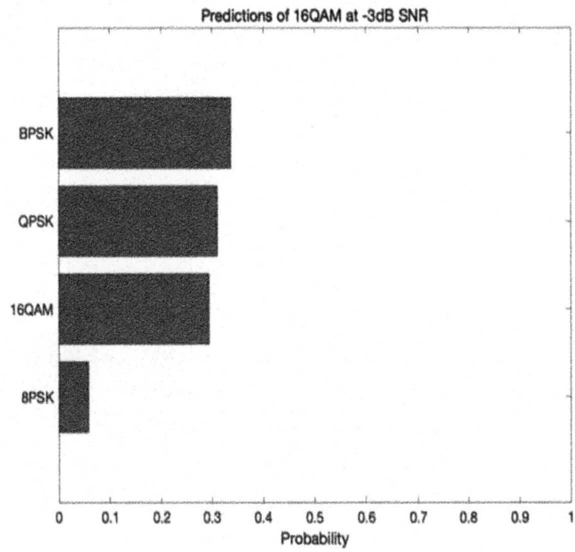

Fig. 5.4. Predictions of 16QAM at −3 dB SNR

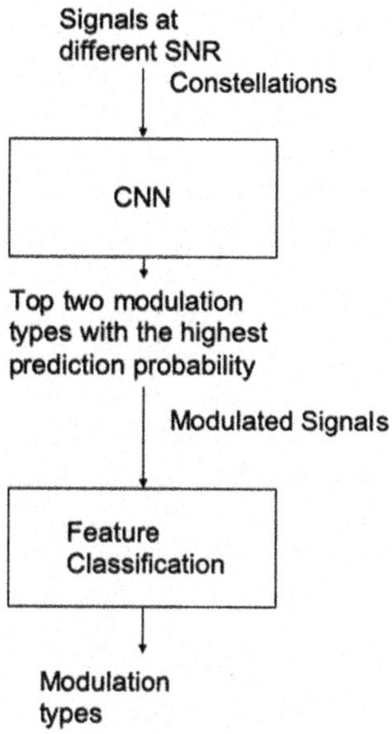

Fig. 5.5 Structure of hybrid model

5.4 Results

This section indicates the results from different classification methods. The Inception-v3 model approves the best results in the deep learning part. It shows the transfer learning can perform better with the pretrained coefficients. To improve the performance of the system, the hybrid model will use the features from the statistical features and be compared to the deep learning model.

5.4.1 *Deep Learning for Images Classification*

Form Table 5.1, the results are shown from the four CNN models. The CNN developed with Iris case performs worse than traditional machine learning methods. It cannot indicate the CNN is unsuitable for detection because the structures and coefficients of the CNN can influence a lot in the classification system. The other three

Table 5.1 Validation accuracy of deep learning models

Model name	Validation accuracy
CNN from Iris case	36.90%
SqueezeNet	55.83%
GoogleNet	56.11%
Inception-v3	63.33%

models are CNN as well, but they have more complicated structures and they are pretrained by millions of images. Although the images applied for pretraining are not the constellation of modulated signals, they can also help the models to classify the images by modifying the coefficient again and again.

The Inception-v3 compared to the SqueezeNet and GoogleNet shows best accuracy of 63.33%. The training dataset is the images of four modulated signals from −10 dB SNR to 20 dB SNR. And the Accuracy is calculated from detect the four modulation types from −10 dB SNR to 20 dB SNR.

Figure 5.6 is the examples for the QAM16 detection at 0 dB SNR and 5 dB SNR by the simple CNN model, SqueezeNet, GoogleNet and Inception-v3 model. The validation accuracy of Inception-v3 is only higher than the CNN model based on the Iris case by 10%. The CNN model developed from Iris model shows a good classification ability at 0 dB SNR, but it cannot promise accuracy at 5 dB SNR. For the higher SNR, the Inception-v3 have a robust classification capability. Considering the validation accuracy and stability of the classification at the same time, we decided to choose the Inception-v3 as the most outstanding one from the four models.

When the SNR is more than 10 dB, the predictions of the classification goes nearly 100%. Deep learning has shown its advantages by the results. Based on these, to improve the classification accuracy of low SNR area, we use the feature classification as a supplement.

5.4.2 Hybrid Model for Images Classification and Feature Classification

According to the Sect. 5.4.1, we find the Inception-v3 provides the best results from the four kinds of CNN structures over all SNR. From Fig. 5.2 we can find the high order cumulants perform well. The high order cumulants are employed for the hybrid system as the second classification process. The Kurtosis and Skewness can describe the shape of the signals. But as the training data, the signals were varied from −10 dB to 20 dB SNR, which means the features also changed dramatically. The dynamic changes will increase the complexity of the classification system. So only C_{20}, C_{21}, C_{40}, C_{41}, C_{42}, C_{63}, C_{80} are employed.

Table 5.2 shows the compared results of Inception-v3 and hybrid model. The accuracy was improved especially at low SNR area. The hybrid model explores signals at the low SNR area and the accuracy is more than 60% by training the whole SNR range data. The accuracy at −5 dB SNR is higher than the accuracy at −3 dB SNR because we capture the constellation images by the scale of the axis from −4 to 4 and the images are full of the random noise pixels compared to the constellations of other modulation types. At the −3 dB SNR, all the four kinds of modulation types have lots of noisy pixels around the focus on the middle part of the constellation, which is more complicated to classify at this period.

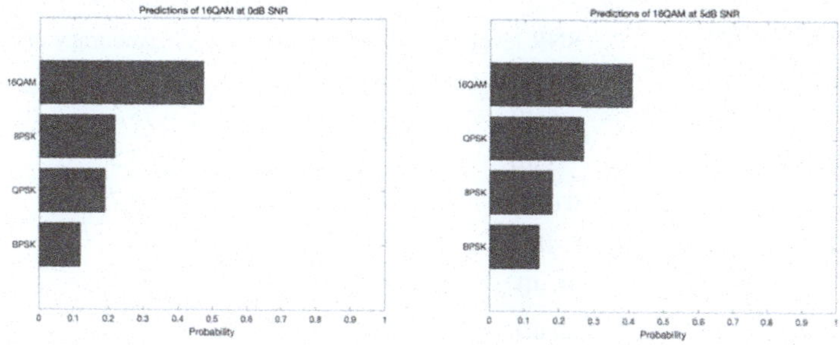

(1) Predictions based on CNN from Iris model.

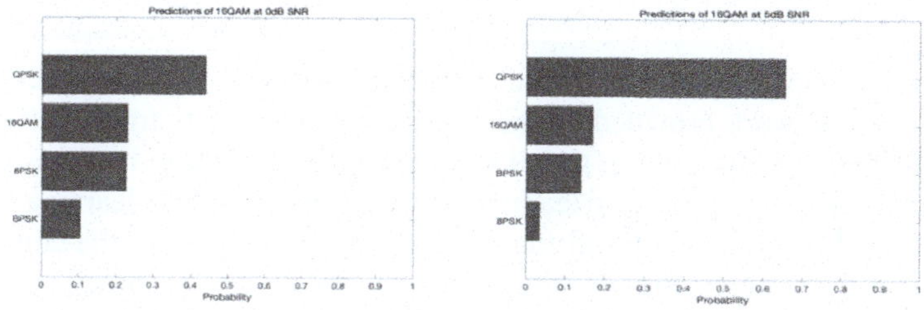

(2) Predictions based on SqueezeNet.

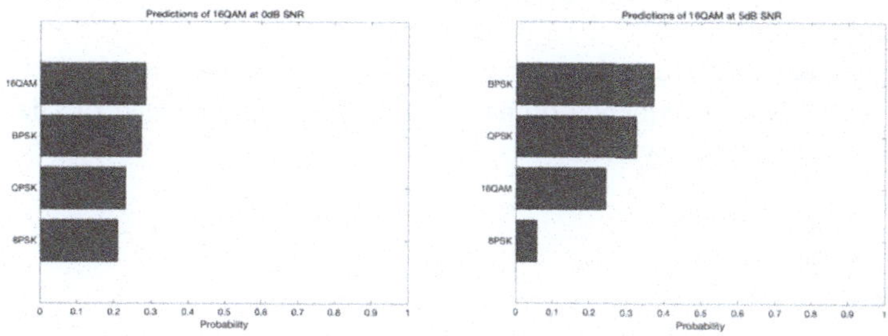

(3) Predictions based on GoogleNet.

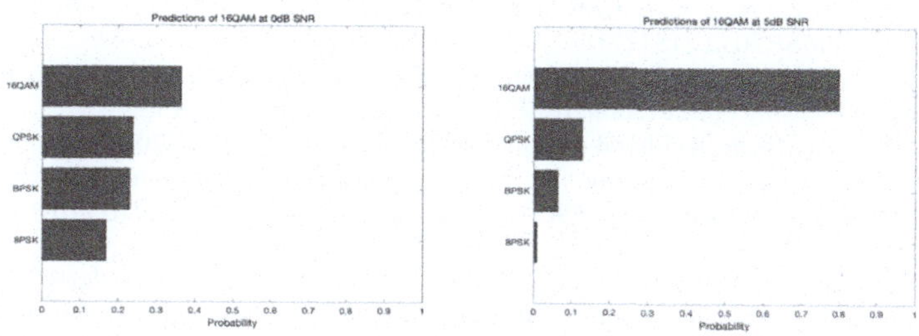

(4) Predictions based on Inception-v3.

Fig. 5.6 a–d Prediction of four deep learning models

Table 5.2 Compared accuracy between CNN and hybrid model

SNR level	Accuracy of Inception-v3 (%)	Accuracy of hybrid model (%)
−5 dB	52.6	75.3
−3 dB	29.5	63.5
0 dB	37.5	67.7
5 dB	79.0	88.5
10 dB	96.7	97.3
20 dB	99.2	99.7

5.5 Conclusion

This work compared the different classification methods for AMC, finding the Inception-v3 network provided the best results at SNR levels from −10 dB SNR to 20 dB SNR. To improve the situation at low SNR, the hybrid model is provided based on the advantages of deep learning and traditional machine learning. The new hybrid model involved the novel image classification to classify the constellation at first and the features classification as a supplement. The accuracy provided by the hybrid model at −3 dB SNR increase from 29.5% to 63.5% compared to the Inception-v3. In this way, the hybrid model utilized the constellation to avoid the complex calculation and several features to improve the performance of previous image classification.

The deep learning technology could still be improved further by detecting the images of the Time–Frequency domain at the same time instead of changing the coefficients of the CNN system.

References

1. Axell, E., Leus, G., Larsson, E.G., Poor, H.V.: Spectrum sensing for cognitive radio: state-of-the-art and recent advances. IEEE Signal Process. Mag. **29**(3), 101–116 (2012). https://doi.org/10.1109/MSP.2012.2183771
2. Hamid, M., Ben Slimane, S., Van Moer, W., Björsell, N.: Spectrum sensing challenges: blind sensing and sensing optimization. IEEE Instrum. Meas. Mag. (2016)
3. Sills, J.A.: Maximum-likelihood modulation classification, pp. 217–220 (1999)
4. Gang, H., Jiandong, L., Donghua, L.: Study of modulation recognition based on HOCs and SVM. IEEE Veh. Technol. Conf. **59**(2), 898–902 (2004). https://doi.org/10.1109/vetecs.2004.1388960
5. Abdelmutalab, A., Assaleh, K., El-Tarhuni, M.: Automatic modulation classification based on high order cumulants and hierarchical polynomial classifiers. Phys. Commun. **21**, 10–18 (2016). https://doi.org/10.1016/j.phycom.2016.08.001
6. Whelchel, J.E., McNeill, D.L., Hughes, R.D., Loos, M.M.: Signal understanding: an artificial intelligence approach to modulation classification, pp. 231–236 (1989). https://doi.org/10.1142/9789814354707_0021

7. Zhang, M., Diao, M., Guo, L.: Convolutional neural networks for automatic cognitive radio waveform recognition. IEEE Access **5**, 11074–11082 (2017). https://doi.org/10.1109/ACCESS.2017.2716191
8. An, N., Li, B., Huang, M.: Modulation classification of higher order MQAM signals using mixed-order moments and fisher criterion. In: 2010 2nd Int. Conf. Comput. Autom. Eng. ICCAE, vol. 3, pp. 150–153 (2010). https://doi.org/10.1109/ICCAE.2010.5451214
9. Spooner, C.M.: On the utility of sixth-order cyclic cumulants for RF signal classification. In: Conf. Rec. Asilomar Conf. Signals, Syst. Comput., vol. 1, pp. 890–897 (2001). https://doi.org/10.1109/ACSSC.2001.987051
10. Dobre, O.A., Bar-Ness, Y., Su, W.: Higher-order cyclic cumulants for high order modulation classification. In: Proc. IEEE Mil. Commun. Conf. MILCOM, vol. 1, C edn, pp. 112–117 (2003). https://doi.org/10.1109/milcom.2003.1290087
11. Lee, J., Kim, B., Kim, J., Yoon, D., Choi, J.W.: Deep neural network-based blind modulation classification for fading channels. In: Int. Conf. Inf. Commun. Technol. Converg. ICT Converg. Technol. Lead. Fourth Ind. Revolution, ICTC 2017, vol. 2017-December, pp. 551–554 (2017). https://doi.org/10.1109/ICTC.2017.8191038
12. Grm, K., Štruc, V., Artiges, A., Caron, M., Ekenel, H.K.: Strengths and weaknesses of deep learning models for face recognition against image degradations. In: IET Biometrics, vol. 7, 1st edn, pp. 81–89 (2018). https://doi.org/10.1049/iet-bmt.2017.0083

Chapter 6
Borderless Table Detection and Extraction in Scanned Documents

Nakul Aggarwal, Snehlata Sinha, Abhilasha Lodha, Debasmita Ghosh, Shalini Singh, and Dharmendra Chaurasia

Abstract Automated detection and extraction of tabular data from scanned documents is a difficult task. Detecting and extracting information from tables without any borders becomes more challenging even with the state-of-the-art OpenCV based solutions. There have been few approaches suggesting the solutions based on deep learning models but none of the open-source solutions performed well when table boundaries are not visible. In this paper, we are proposing a novel end-to-end solution that works well with borderless tables of different layouts and templates. The proposed method consists of two stages. First stage is automated table detection irrespective of the format, border and layout of the scanned documents using custom trained deep learning model. Second stage is a novel OpenCV based custom algorithm which utilize blob detection to define row–column relationship of each cell. This information is used to extract the data into.csv/.xlsx format. Our proposed algorithm was evaluated on more than 100 images and has achieved an average table extraction accuracy of 98.4% outperforming many state-of-the-art models.

Keywords Borderless table detection · Table extraction · OpenCV · Deep learning · Blob detection

6.1 Introduction

Human society in today's world has an ever-increasing wealth of information with millions of paper documents being produced every day. Tables are extensively used for exhibiting structural and functional information [1]. They are effective and compact means of displaying information such as numeric values [2] to human readers as they make such information easy to search, compare and understand. Therefore, table detection from documents is considered an essential part of a document analysis through the extraction of summarized and well-organized information [3]. However, extraction of tabular data in scanned documents using

---

N. Aggarwal (✉) · S. Sinha · A. Lodha · D. Ghosh · S. Singh · D. Chaurasia
Advance Technology Centers in India, Accenture, India
e-mail: nakul.aggarwal@accenture.com

M. A. Laribi et al. (eds.), *Advances in Automation, Mechanical and Design Engineering*, Mechanisms and Machine Science 121,
https://doi.org/10.1007/978-3-031-09909-0_6

Optical Character Recognition (OCR) is a challenging task due to the presence of a wide variety of table formats and layouts.

A lot of OpenCV and OCR based solutions have been proposed for tables with visible lines and borders. However, they are not able to perform well on the borderless tables i.e., tables without horizontal and vertical lines for demarcation. Our paper proposes an end-to-end pipeline which takes an input scanned document and extracts the tabular data in an editable format (.csv/.xlsx). We have collected borderless table dataset of 300 images, 150 from TableBank [4], 75 from internet and 75 are custom scanned invoice images. This data contains variations like different backgrounds, brightness, angles, etc. to accommodate real-life scenarios.

The input image is pre-processed and sent for table detection and extraction. For table detection, a custom trained deep learning model is used to detect the tables irrespective of borders, format or layout. For table extraction, a novel algorithm is designed to define the row column relationship of each cell as depicted in the Sect. 6.3. Tesseract [5] is used to extract raw data from the identified cells. The detected cell ROI are pasted on an empty image to decrease the amount of time it takes for Tesseract extraction as explained further in section III-E. Once cell extracts are retrieved, they are arranged back in the same tabular structure to the spreadsheets.

In order to check the performance of our solution, the parameters taken into consideration are time taken for extraction, precision, recall and F1-Score as shown in Sect. 6.4.

6.2 Literature Review

Jihu Kim and Hyoseok Hwang [6] presented a rule-based method for detecting tables in website images. They proposed a pipeline in two stages, feature extraction based on constraint rules followed by grid pattern recognition using rules and tree structures. They were able to reduce number of false positives for non-text blocks in images. Their method outperforms previous methods for website images. However, the performance is not good for document images.

Sahoo et al. [1] have proposed the Auto-Table-Extract system to automatically detect and extract information from the table in a pdf file into an excel file. It performs well on scanned and searchable files. However, performance is not good for partially bordered and borderless tables.

Prasad et al. [7] proposed instance segmentation-based Cascade TabNet to recognize structures within tables by predicting table cell masks while using the line information as well. It performs good on public datasets and suggests improving the post-processing modules for increasing accuracy.

Fisher et al. [8] presented a multistage pipeline for table detection and table structure recognition with document alignment and color invariance pre-processing methods. Their solution does not generalize for complex tables with cells or entire table within a cell.

Shafait and Smith [9] proposed an approach using a layout analysis of Tesseract [17] with tab-stops used as an indication of where a text block starts and ends. Their table detection approach competes well with many commercial OCR systems. The detection accuracy of their algorithm for the UNLV dataset achieved a precision of 86% and a recall of 79%.

Hassan et al. [10] detect tables in PDF documents utilizing both ruling lines and content layout. However, their method utilizes these two sources separately. And lots of false positive tables are detected because the detected graphic lines are not verified first in their method.

Similarly, Fang et al. [11] propose a method via both visual separators and tabular structures of contents. The separators refer to not only graphic lines but also white spaces to handle borderless tables. This method detects page columns in the first place to assist table detection in multi-column pages and achieves a satisfactory accuracy rate.

Liu et al. [12, 13] proposed a method in PDF based on the notice that almost all the table rows are sparse lines. Therefore, the table boundary detection problem could be simplified into the sparse line analysis problem with much less noise. The authors designed eight-line label types and applied two machine learning techniques, Conditional Random Field (CRF) and Support Vector Machines (SVM), on the table boundary detection field.

6.3 Methodology

Our solution offers an end-to-end flow for borderless table detection and extraction. Figure 6.1 illustrates the pipeline for the same. The document uploaded is passed to pdf to jpg module in case it's a pdf. The input images are then passed to the skewness correction module. This is followed by a deep learning-based object detection module to detect bordered and borderless table layouts present in the image. In case of images consisting of borderless tables, the detected table crop is then pre-processed using OpenCV techniques and passed for detection of cells in the table. The cells

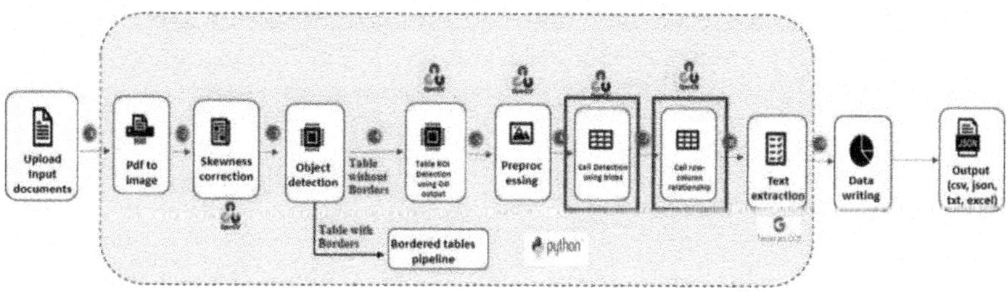

Fig. 6.1 Borderless table detection and extraction pipeline

are detected using blob-detection technique which is then followed by cell row–column relationship module to define the position of cells in rows and columns. Text extraction module extracts the data present in table cells followed by data writing to support output formats. Data writing module has the capability of writing the detected table extracts to a spreadsheet table which is a CSV file. All these different modules are packed together to create the borderless table detection and extraction pipeline.

6.3.1 *Data Acquisition*

Our research is focused on extracting borderless tables of both uniform and non-uniform types. Our model consists of 2 classes, bordered and borderless tables as shown in Fig. 6.2. Around 300 images were collected, 150 per class from Table-Bank, 75 from internet and 75 are custom scanned invoice images. The dataset has varied angles, layout, background and quality to accommodate multiple variations. The dataset included invoices, scanned contract document images, receipts etc. The input images are in .jpg, .png and .jpeg format. The images are then annotated for tables using the annotation tool LabelImg [14] and json files gets created for each of them.

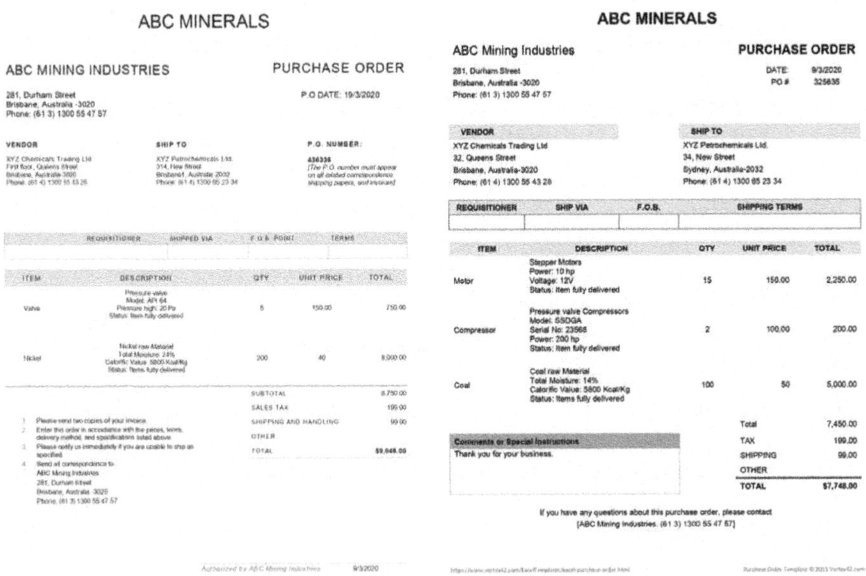

(a) (b)

Fig. 6.2 Dataset sourced from Tablebank, custom and internet

6.3.2 *Object Detection Model for Table Detection*

An object detection model is trained on custom dataset to identify bordered and borderless tables in the image. The architecture used is SSD Mobilenet V1 FPN [15] pretrained on coco dataset [16]. The trained model was converted to TensorFlow Lite (TF Lite) [17] model for improved performance.

The main advantage of using the SSD architecture is its ability to detect multiple objects in the image with a high MAP score. It also provides best accuracy tradeoffs within the fastest detectors.

In our pipeline, the input image is initially checked for skewness. An OpenCV based module is used to calculate and adjust the angle of the input image. The skew-corrected image is then passed through the custom trained object detection TF Lite model. The custom model was trained using x2gd.xlarge ec2 instance with 4 vCPU and 64 GB memory. If the class of detected table in the image is'bordered table', it is passed to the bordered table pipeline. If the class of detected table is'borderless table', coordinates of the detected table are used to extract the table region from the image. A table crop is generated and sent for further processing to extract table data as explained in Sect. 6.3.3.

6.3.3 *Image Preprocessing*

To handle table with varied backgrounds, adaptive thresholding was done on the input image. To get better cell detection, any horizontal or vertical lines present in the image were removed using the line removal module. The cropped table ROI detected image was further passed to the pre-processing stage i.e., conversion to gray scale image by applying denoising filters. The resultant image after pre-processing is shown in Figs. 6.3 and 6.4.

6.3.4 *Cell ROI Detection and Extraction*

After the table image was pre-processed, cells were detected using the OpenCV based cell blob detection technique. A blob is a group of connected pixels that share a common property and is widely used technique in OpenCV based applications.

Cells blobs were detected by doing contour detection on the vertically dilated image as shown in the Fig. 6.5 and detected cells were shown in Fig. 6.6 on table image.

To detect the position of the cells, a list of the coordinates (x,y,w,h) of the contours was extracted and were sorted from top-to-bottom. These coordinates were used to define the row–column relationship of each cell in the table with accurate starting and ending row–column cell information.

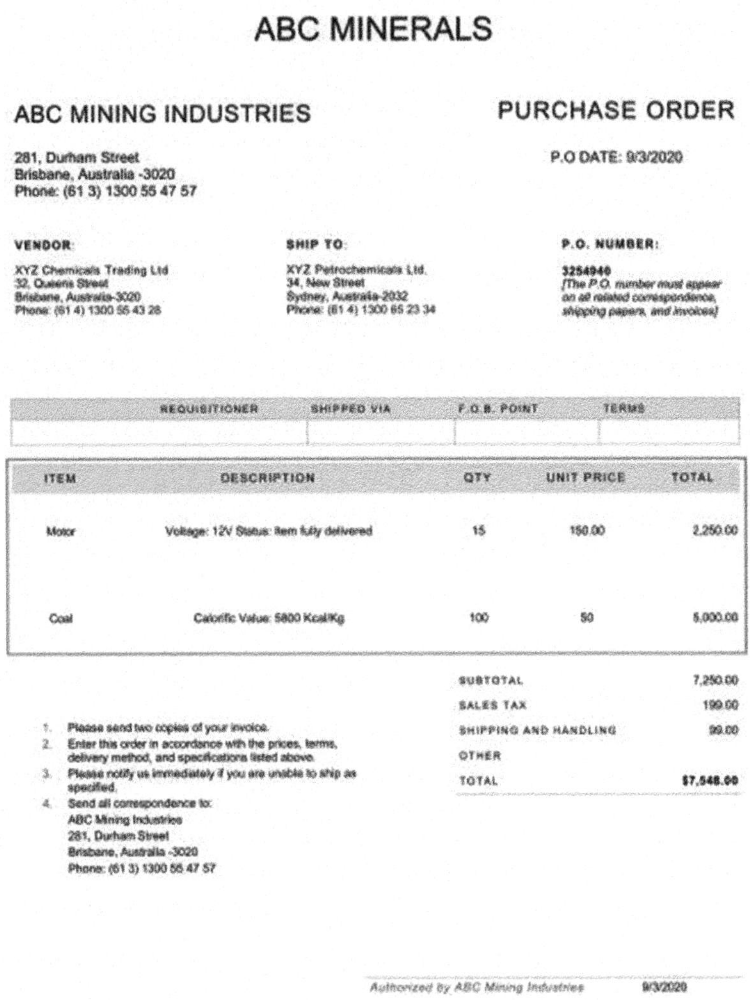

ABC MINERALS

ABC MINING INDUSTRIES

PURCHASE ORDER

281, Durham Street
Brisbane, Australia -3020
Phone: (61 3) 1300 55 47 57

P.O DATE: 9/3/2020

VENDOR:
XYZ Chemicals Trading Ltd
32, Queens Street
Brisbane, Australia-3020
Phone: (61 4) 1300 55 43 28

SHIP TO:
XYZ Petrochemicals Ltd.
34, New Street
Sydney, Australia-2032
Phone: (61 4) 1300 65 23 34

P.O. NUMBER:
3254940
[The P.O. number must appear on all related correspondence, shipping papers, and invoices]

REQUISITIONER	SHIPPED VIA	F.O.B. POINT	TERMS

ITEM	DESCRIPTION	QTY	UNIT PRICE	TOTAL
Motor	Voltage: 12V Status: Item fully delivered	15	150.00	2,250.00
Coal	Calorific Value: 5800 Kcal/Kg	100	50	5,000.00

SUBTOTAL	7,250.00
SALES TAX	199.00
SHIPPING AND HANDLING	99.00
OTHER	
TOTAL	$7,548.00

1. Please send two copies of your invoice.
2. Enter this order in accordance with the prices, terms, delivery method, and specifications listed above.
3. Please notify us immediately if you are unable to ship as specified.
4. Send all correspondence to:
   ABC Mining Industries
   281, Durham Street
   Brisbane, Australia -3020
   Phone: (61 3) 1300 55 47 57

Authorized by ABC Mining Industries 9/3/2020

Fig. 6.3 Table detection in the input image

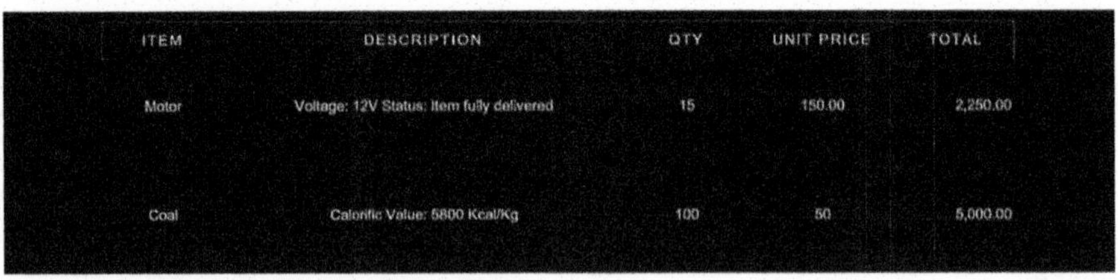

ITEM	DESCRIPTION	QTY	UNIT PRICE	TOTAL
Motor	Voltage: 12V Status: Item fully delivered	15	150.00	2,250.00
Coal	Calorific Value: 5800 Kcal/Kg	100	50	5,000.00

Fig. 6.4 Processed image after adaptive thresholding

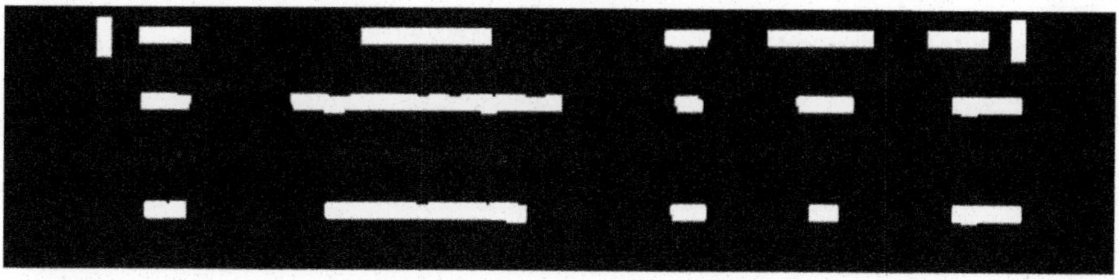

Fig. 6.5 Horizontally and vertically dilated cells

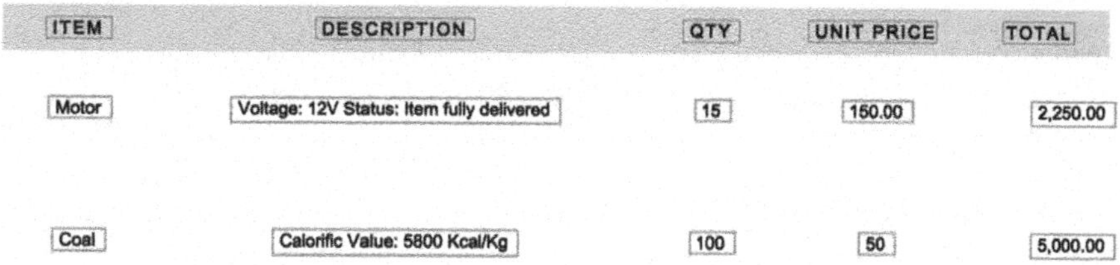

Fig. 6.6 Cells detected in the table

Our custom algorithm can be used to extract data from all types of borderless tables. Here we used minimum x and y values, to trace along the table in ascending order to detect each row and column. The row–column relationship for each cell is defined using five parameters namely cell coordinates, start row number, end row number, start column number and end column number. To check if the table is uniform or non-uniform, we also calculated average height of cells in the same row. If cell height is less than or equal to average height, then table is tagged as a uniform borderless table, else it is tagged as a non-uniform borderless table.

In case of non-uniform tables, we used maximum y values of merged cell and traced along the table until its value matched with minimum y value of another cell. Once done, the end row number of the merged cell would be equal to the start row number of the identified cell.

Finally, a refined table content was extracted, and the cell row–column information was sent to a csv writer module to interpret the formation and provide a csv/json output as shown in Fig. 6.7.

0	1	2	3	4
ITEM	DESCRIPTION	QTY	UNIT PRIC	TOTAL
Motor	Voltage: 12V Status: Item fully delivered	15	150.00	2,250.00
Coal	Calorific Value: 5800 Kcal/Kg	100	50	5,000.00

Fig. 6.7 Extracted table data in CSV format

(a) (b)

Fig. 6.8 **a** Round 1 with white background. **b** Round 2 with input ROI background color

6.3.5 Performance Optimization

Reduction of pyTesseract API calls using OpenCV Techniques—OpenCV processing is known for being very fast with images. When cell ROI crops are detected, each crop image is then converted into text using Tesseract OCR. The pyTesseract API takes about a second approximately when each cell ROI crop are sent to it. This way, the computational time increased in the cell extraction module. In order to resolve this issue, a new approach is introduced in this paper.

A background mask is created by calculating the height and width of all cell ROI crops. In in the background mask, all the crops are arranged one below the other, having a delimiter in between, and then the image is sent to pyTesseract API for the text extraction. Figure 6.8 shows the way the ROI crops were arranged with delimiter ROI. Figure 6.8a shows the first round of implementation where we used a white background mask for arranging the ROI crops and then sending it for Tesseract extraction. With this approach, as the ROI crop background wasn't necessarily being white in color all the time, the OCR extraction had some faults. Then in Round 2 (shown in Fig. 6.8b) the ROI crop's background pixel was detected and then that pixel was used as a background mask color. This reduced the OCR faults while doing text extraction.

The cell data which is extracted from Tesseract, is segregated back to rows and columns using the delimiter and coordinates or the ROI crops. This approach reduces the time taken by the pyTesseract API to extract the textual data out of the tables.

Table 6.1 describes the time differences in processing text extraction from ROI crops using the default approach of sending each ROI crop for text extraction vs sending one image with arranged ROI crops.

Table 6.1 Time taken for Table Data Extraction once cell ROIs are detected

Table	Basic approach	Background mask approach
Table with 24 cells	24 s	3 s
Table with 24 cells	50 s	4 s

6.4 Result and Conclusion

Our proposed solution offers multiple novelties over the other state of the art algorithms. Being a two-step approach, our solution aids in accurate table detection and row and column data extraction.

At the first step, our custom trained object detection module identifies borderless tables in scanned input documents explicitly. Then at the next step, an OpenCV based novel algorithm identifies cells (rows and columns) from the table extracted in the previous step. This algorithm defines row–column relationship for each cell of the table. Accuracies for both steps are mentioned in Table 6.2.

The performance of our system is evaluated based on the three model evaluation metrics termed as precision, recall and F1 score.

A. Precision—Precision can be evaluated by finding out the number of cell crops that were correctly identified to the total number of identified cell crops by our OpenCV module.

$$\text{Precision} = \frac{\text{Number of correctly identified cell crops}}{\text{Total number of identified cell crops}} \tag{6.1}$$

B. Recall—Recall can be evaluated by finding the number of cell crops that were correctly identified by our OpenCV module to the total number of cell crops present.

$$\text{Recall} = \frac{\text{Number of correctly identified cell crops}}{\text{Total number of cell crops in the document}} \tag{6.2}$$

C. F1-score—F1-score can be evaluated with the help of precision and recall which were calculated in the Eqs. (1) and (2).

$$\text{F1} = \text{F1} - \text{score} = 2 * \frac{(\text{Precision} * \text{Recall})}{(\text{Precision} + \text{Recall})} \tag{6.3}$$

The Fig. 6.3 represents accurate detection of borderless tables from scanned documents by our custom trained object detection module. The precise identification and row and column data extraction from the detected table is depicted in Figs. 6.5, 6.6 and 6.7.

In order to identify the effectiveness of our solution, we have evaluated our 2-step method with a dataset comprising of more than 100 PDFs irrespective of the

Table 6.2 Accuracy

Table Detection Accuracy	96.4%
Data Extraction Accuracy	98.8%

Table 6.3 Performance evaluation

Type of table	Precision	Recall	FL-score
Bordered table	0.97	1.00	0.98
Borderless table	0.95	0.98	0.96

format, layout and border of the scanned documents. Table 6.3 represents the model evaluation metrics (precision, recall and F1 score) used for performance evaluation of our novel solution.

6.5 Future Works

The performance of our solution is comparable with some state-of-the-art commercial and open-source tools available in the market and overcomes the limitations offered by them.

However, there are certain challenges offered by our solution that we would inculcate in our future works. Firstly, our solution can't process handwritten text. Also, our object detection module used to extract tables out of images can be improvised by training the model on more train data. Lastly, the horizontal and vertical dilation-erosion techniques can be automated.

References

1. Sahoo, R., Kathale, C., Kubal, M., Malik, S.: Auto-table-extract: a system to identify and extract tables from pdf to excel, vol. 9, pp. 217–221 (2020)
2. Ohta, M., Yamada, R., Kanazawa, T., Takasu, A.: A cell-detection-based table-structure recognition method. Association for Computing Machinery, New York, NY, USA (2019). https://doi.org/10.1145/3342558.3345412
3. Hashmi, K.A., Liwicki, M., Stricker, D., Afzal, M., Afzal, M., Afzal, M.Z.: IEEE Access, vol. PP, pp. 1–1 (2021)
4. Li, M., Cui, L., Huang, S., Wei, F., Zhou, M., Li, Z.: Tablebank: Table benchmark for image-based table detection and recognition. CoRR, vol. abs/1903.01949, 2019. http://arxiv.org/abs/1903.01949
5. Smith, R.: An overview of the tesseract ocr engine. In: Nineth Inter-national Conference on Document Analysis and Recognition (ICDAR 2007), vol. 2, pp. 629–633 (2007)
6. Kim, J., Hwang, H.: A rule-based method for table detection in website images. IEEE Access, vol. 8, pp. 81 022–81 033 (2020)
7. Prasad, D., Gadpal, A., Kapadni, K., Visave, M., Sultanpure, K.: Cascadetabnet: An approach for end to end table detection and structure recognition from image-based documents. *CoRR*, vol. abs/2004.12629 (2020)
8. Fischer, P., Smajic, A., Mehler, A., Abrami, G.: Multi-type-td-tsr—extracting tables from document images using a multi-stage pipeline for table detection and table structure recognition: from OCR to structured table representations. *CoRR* vol. abs/2105.11021 (2021)
9. Shafait, F., Smith, R.: Table detection in heterogeneous documents, vol. 01, pp. 65–72 (2010)

10. Hassan, T., Baumgartner, R.: Table recognition and understanding from pdf files. In: Proceedings of the Ninth International Conference on Document Analysis and Recognition, vol. 02, ser. ICDAR '07, pp. 1143–1147. IEEE Computer Society, USA (2007)
11. Fang, J., Gao, L., Bai, K., Qiu, R., Tao, X., Tang, Z.: A table detectionmethod for multipage pdf documents via visual seperators and tabular structures. In: 2011 International Conference on Document Analysis and Recognition, pp. 779–783 (2011)
12. Liu, Y., Bai, K., Mitra, P., Giles, C.L.: Improving the table boundary detection in pdfs by fixing the sequence error of the sparse lines. In: 2009 10th International Conference on Document Analysis and Recognition, pp. 1006–1010 (2009)
13. Liu, Y., Mitra, P., Giles, C.L.: Identifying table boundaries in digital documents via sparse line detection. In: Proceedings of the 17th ACM Conference on Information and Knowledge Management, ser. CIKM '08, pp. 1311–1320. Association for Computing Machinery, New York, NY, USA (2008). https://doi.org/10.1145/1458082.1458255
14. Tzutalin, "Labelimg," Free Software: MIT License, 2015. https://github.com/tzutalin/labelImg
15. Howard, A.G., Zhu, M., Chen, B., Kalenichenko, D., Wang, W., Weyand, T., Andreetto, M., Adam, H.: Mobilenets: efficient convolutional neural networks for mobile vision applications. *CoRR*, vol. abs/1704.04861 (2017). http://arxiv.org/abs/1704.04861
16. Lin T.-Y., Maire, M., Belongie, S., Bourdev, L., Girshick, R., Hays, J., Perona, P., Ramanan, D., Zitnick, C.L., Dolla´r, P.: Microsoft coco: Common objects in context (2015)
17. Abadi, M., Agarwal, A., Barham, P., Brevdo, E., Chen, Z., Citro, C., Corrado, G.S., Davis, A., Dean, J., Devin, M., Ghemawat, S., Goodfellow, I., Harp, A., Irving, G., Isard, M., Jia, Y., Jozefowicz, R., Kaiser, L., Kudlur, M., Levenberg, J., Mane´, D., Monga, R., Moore, S., Murray, D., Olah, C., Schuster, M., Shlens, J., Steiner, B., Sutskever, I., Talwar, K., Tucker, P., Vanhoucke, V., Vasudevan, V., Vie´gas, F., Vinyals, O., Warden, P., Wattenberg, M., Wicke, M., Yu, Y., Zheng, X.: TensorFlow: Large-scale machine learning on heterogeneous systems (2015). software available from tensorflow.org. https://www.tensorflow.org/

Chapter 7
Floor Space Optimisation and Recommendation System in 2D Space

Nakul Aggarwal, Asmita Mahajan, Snehlata Sinha, Abhilasha Lodha, Debasmita Ghosh, and Vinay Dilip Raut

Abstract In this paper, we are proposing an open-source solution for empty space calculation in images by segmenting out the objects using computer-vision technologies. We have pre-existing solutions for 3D floor mapping but there is no end-to-end approach that can be implemented in a 2D space. Our solution takes an input image of a scene like an empty balcony or a room and provides us a 2D floor mapping of the space. Using transfer learning, a custom trained semantic segmentation model is used to identify the objects in images. The outcome of semantic segmentation model is used by our custom algorithm to determine the empty floor space in an image and segment the area into restricted and non-restricted regions. The restricted segments, like the pathway in-front of the doors, furniture etc., are the ones that should be kept empty for people to move around easily. The non-restricted floor area can be utilised for a wide number of use-cases like design recommendations, autonomous robot navigation, empty parking space identification etc. This paper mainly deals with the balcony garden design recommendation, however as mentioned the same approach can be extended to other spaces.

Keywords Computer vision · 2D space · Floor mapping · Transfer learning · Recommendation system

7.1 Introduction

Efficient usage of floor space mapping and recommendation are extremely desired by house planners, interior decorators, architects, etc. to suffice functional as well as artistic purposes. Creating floor-maps is a creative process that requires imagination as well as mathematical skills. People can be trained to perform the task relatively well; however, hiring a professional workforce can be quite expensive and time-consuming. This can be best achieved by an Artificial Intelligence based

---

N. Aggarwal (✉) · A. Mahajan · S. Sinha · A. Lodha · D. Ghosh · V. D. Raut
Advance Technology Centers in India, Accenture, India
e-mail: nakul.aggarwal@accenture.com

M. A. Laribi et al. (eds.), *Advances in Automation, Mechanical and Design Engineering*, Mechanisms and Machine Science 121,
https://doi.org/10.1007/978-3-031-09909-0_7

recommendation system that provides a cost-effective way of obtaining the otherwise expensive space-optimization services.

Recently, number of solutions that employ floor mapping in 3D space have been implemented using Extended Reality (XR) and Augmented Reality (AR). But there are no appropriate end-to-end solutions readily available that can implement floor mapping in a 2D space. Here, we are proposing a highly optimized open-source solution which takes an input image of an empty balcony and gives out design recommendations for placement of plants in the detected empty regions. So, our solution offers an end to end approach for empty floor space calculation in 2D spaces by segmenting out the objects using computer vision based semantic segmentation algorithm.

The key benefit of using semantic segmentation over object detection is that it identifies objects present in an image at the pixel level [1]. This helps in precise localisation of the empty floor space area and thus aids in an efficient floor area calculation. The model architecture used in developing the solution is Mask R-CNN. Using transfer learning, this model is trained on custom dataset to segment six classes namely wall, door, window, ceiling, railing and furniture present inside the image. A custom logic is developed to determine the empty floor space in the image using the output of semantic segmentation model. The calculated empty floor space is then segmented into restricted and non-restricted regions.

The restricted area is the one that is kept empty for people to move around easily, like pathway in front of doors, furniture etc. The non-restricted floor area can be effectively utilised for wide number of applications like indoor design recommendations, autonomous robot navigation, empty parking space identification etc. Here we have demonstrated a similar use case for balcony garden design recommendation. For every identified non restricted area, the number of plant rows are calculated and using Image Blending technique, the plant images get seamlessly blended into the non-restricted segments, thereby giving a beautiful balcony garden design recommendation.

7.2 Literature Review

7.2.1 Related Architectures

Cheng et al. [2], proposed the current cutting edge semantic segmentation approach based on DeepLabV3+, who enhanced DeepLabV3 [3] into a encoder-decoder model. The original FCN (2015) had a mean IOU of 67.2% and DeepLabV3+ (2018) has IOU of 67.2% on the PASCAL VOC 2012 dataset. DeepLabV3+ outperforms many SOTA approaches like ResNet-38 and PSPNet with an IOU of 87.83% on the PASCAL VOC 2012 dataset with DeepLabV3+ (Xception) method and an IOU of 89.00% with DeepLabV3+ (Xception-JFT) method as it works effectively in detecting the object boundaries. However, architectural properties can affect the robustness of a model significantly. The generalization capability of DeepLabv3+ model, using a ResNet-backbone, depends strongly on the type of image corruption.

Wang et al. [1] have utilised deep Convolutional Neural Networks (CNNs) to get a significant improvement over previous semantic segmentation systems outperforming FCN, DilatedNet, and DeepLabv2. They have shown how to improve pixel-wise semantic segmentation by manipulating convolution related operations. They have tested their solution on Cityscapes, KITTI road estimation and PASCAL VOC2012 dataset. Their model, ResNet-DUC-HDC, have been designed with a new dense up sampling convolution (DUC) operation to enable pixel-level prediction on feature maps, and hybrid dilated convolution (HDC) to solve the gridding problem, effectively enlarging the receptive fields of the network. However, there are SOTA models like DeepLabV3+ and Mask-RCNN that has outperformed ResNet-DUC-HDC in certain scenarios.

Girshick [4] proposed a Fast Region-based Convolutional Network method (Fast R-CNN) for object detection. Fast R-CNN builds on previous work to efficiently classify object proposals using deep convolutional networks. Compared to previous work, Fast R-CNN employs several innovations to improve training and testing speed while also increasing detection accuracy. Then Ren et al. [5] merged Region Proposal Network (RPN) and Fast R-CNN into a single network by sharing their convolutional features using the recently popular terminology of neural networks with 'attention' mechanisms. They were able to achieve state-of-the-art object detection accuracy on PASCAL VOC 2007, 2012, and MS COCO datasets with only 300 proposals per image. Further He et al. [6] proposed an approach for efficient detection of objects in an image while simultaneously generating a high-quality segmentation mask for each instance. The method is called Mask R-CNN, it extends Faster R-CNN by adding a branch for predicting an object mask in parallel with the existing branch for bounding box recognition. Mask R-CNN outperforms all the existing, single-model entries on every task, including the COCO 2016 challenge winners. Abdulla [7] keeping [6] as baseline he has utilised Mask R-CNN for object detection and segmentation. His model generates bounding boxes and segmentation masks for each instance of an object in the image. It's based on Feature Pyramid Network (FPN) and ResNet101, trained on coco dataset [8].

7.2.2 Related Datasets

Dodge et al. [9] have used FCN for segmentation of walls, Faster R-CNN for detection of rooms and Googles character recognition API to detect the size of the room. They have introduced R-FP [9], a dataset of the floor plan which labels wall segmentation. They have observed IOU metrics for the wall segmentation which concludes that models trained over one dataset give poor results when tested on other datasets. Liu et al. [10] converted bitmap images into vectors using a pipeline that consists of semantic segmentation, junction prediction, integer programming and vector format processing. They have used 870 self-labelled images, with a training-test split of

88.5%—11.5%. Liu et al. have reported recall and accuracy metrics calculated from IOU for rooms and objects, and euclidean distances for the walls. This paper achieves a significant performance boost with a large corpus of floorplan vector data and popup 3D models, opening potentials for a new family of big-data Computer Vision research.

Kalervo et al. [11], used a similar pipeline as Liu et al. but they have presented a novel image dataset called CubiCasa5K, a large-scale floorplan image dataset containing 5000 samples annotated into over 80 floorplan object categories. They have reported overall and mean accuracy, mean IOU for the segmentation map, and accuracy and recall for the junctions, rooms and objects. Their implementations on the novel dataset have significantly improved the research on automatic floorplan image analysis as it presents a more elaborate means for examining the problem in a comprehensive manner.

Liu et al. [12] have proposed FloorNet, a novel deep neural architecture for automatically reconstructing a floorplan in indoor space. This paper has set a benchmark for floorplan reconstruction by acquiring RGBD video streams for 155 residential houses or apartments and annotating complete floorplan information.

7.3 Methodology

Our solution offers an end to end flow for empty floor space optimization and recommendation system for balcony garden in 2D space. Figure 7.1 illustrates the pipeline for the same. The image uploaded is first passed for quality check, where if the standards are not met, the user is asked to upload the image again. The quality check approved image is then passed to custom trained semantic segmentation model to segment 6 classes namely wall, door, window, ceiling, railing and furniture.

The output from the semantic segmentation model is then used to calculate empty floor space in the image by our custom algorithm. The identified floor area is further segmented to restricted and non-restricted areas marking them as red and green zones

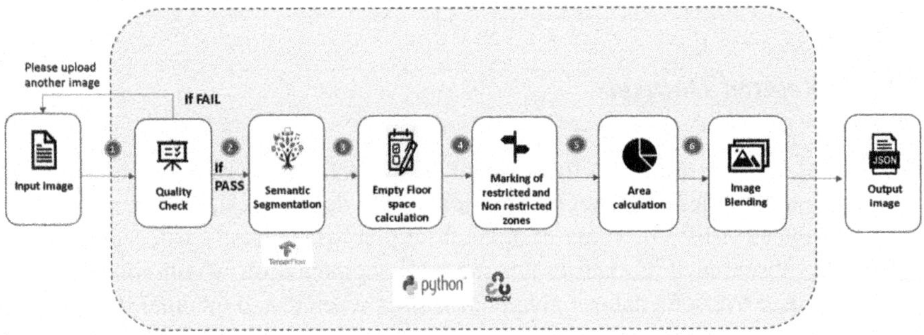

Fig. 7.1 Floor space optimisation and recommendation system pipeline

respectively. Using the non-restricted floor area, the number of rows of plants to be placed is calculated. The plant images are seamlessly blended with the uploaded image background using our image blending module to give a beautiful balcony design recommendation. All these individual modules are packed together to create the end to end pipeline.

7.3.1 Data Acquisition

As we proceed with our algorithm to extract empty floor space, we come across various indoor and outdoor space datasets that suit our purpose. Our research here is primarily focused on dataset consisting of different balcony images. The input images are in .jpg, .png and .jpeg format. The primary source of our dataset are Adobe Photoshop, internet and custom balcony images with varied angles, backgrounds and quality captured by us using different phone cameras as shown in Fig. 7.2.

(a) (b) (c) (d)

Fig. 7.2 Dataset sourced from Adobe photoshop, internet and custom

The balcony dataset images are further annotated for six classes namely wall, door, window, ceiling, railing and furniture for segmentation. An open-source VIG annotation tool is used and JSON files are created for each class mentioned above. The entire dataset is split into 80% train, 10% test and 10% validation set.

7.3.2 *Quality Check*

The input image is first passed to the quality check module that checks the quality of the input image to be processed further. The image quality is checked based on two parameters- brightness and blurriness.

For brightness we calculate the mean of all the pixels in each channel using the Pillow library. This returns the average pixel brightness of three channels. We use a threshold value and if the average is less than the threshold, then the image is less bright. Blurriness is related to the sharpness of an image and a blurry image does not have clear edges. So here, we have used the specific edge detection filter which is the Laplacian filter. It returns the average variance of the edges in an image. The higher the number, the sharper the edge is. We use a threshold value and if Laplacian variance is less than the threshold, we can state that the image is blurry.

If the image passes the set threshold for both these parameters, it is passed on for further processing which is to go through the semantic segmentation module for empty floor space detection in the image. If the threshold parameters are not met and the standard quality check fails, the user is asked to upload a new image with better quality.

7.3.3 *Semantic Segmentation Model*

Once the image has passed the quality check module mentioned in Sect. 7.3.2, it then goes through the semantic segmentation model to identify the six annotated object classes in the image. A semantic segmentation model [13] is trained on our custom data set mentioned in Sect. 7.3.1. Semantic segmentation identifies objects present in an image at the pixel level and aids in precise localisation of the empty floor space area. There are four available architectures to perform semantic segmentation namely R-CNN, Fast R-CNN, Faster R-CNN and Mask R-CNN. Mask R-CNN is an extension to Faster R-CNN and detects multiple objects, objects of different scales, and overlapping objects in an image. This improves the speed and efficiency for object identification and that's the reason we have used Mask R-CNN pre-trained on coco dataset [8].

Architecture and Working of Mask R-CNN.

Mask R-CNN uses Resnet 101 architecture to extract features from the image which act as an input for the next layer. These features are then applied to region proposal

Table 7.1 Hyper-parameters used in model training

Epochs	Steps per epoch	Learning rate	Detection confidence
10,000	1000	0.001	0.9

network (RPN). Here, we get the regions which model predicts contain some object. These regions are then passed to fully connected network to predict class labels and bounding boxes. Additionally, Mask R-CNN generates masks for each class. For all predicted Regions of Interest (RoI), Intersection over Union (IoU) is computed. If IoU is greater than the set threshold (0.5), we consider that RoI, else neglect it. Once we have our ROIs, mask branch is added to the architecture. This returns segmentation mask of size 28 × 28 for each region that contains an object, which is then scaled up for inference.

The hyper-parameters are fine-tuned and the model is trained with the parameters mentioned in Table 7.1. Continuous learning is also integrated with this module. Therefore the model will keep on learning as new data is fed to the model over time. This will enhance the model performance and accuracy. With continuous learning, the semantic segmentation model can also be utilised for other applications like parking space identification, indoor design recommendation etc. The output of the semantic segmentation model (Mask R-CNN) gives the bounding box coordinates of the regions identified in the image. A custom algorithm is developed to determine the empty floor space in the image using the output of semantic segmentation model.

7.3.4 *Empty Floor Space Calculation*

The output of the semantic segmentation model helps us identify the objects present in the input image as shown in the Fig. 7.3. Using this, we try to identify the empty space present in the image. A custom logic is built to identify the four boundary points of the empty floor space using the coordinates of the identified objects.

These coordinates are then used to find the intersection points between them. For example, in the Fig. 7.3, the bottom-right corner of railing is intersecting with bottom-left corner of the identified wall. After getting intersection points, distance between these points is calculated and a polygon is drawn as shown in the Fig. 7.4 below. This polygon depicts the identified empty region.

7.3.5 *Marking of Restricted and Non-Restricted Regions*

Once the empty space is identified as explained in the Sect. 7.3.4, we have to mark the specific region which can be actually utilized further. For this, the entire space is divided into restricted and non-restricted regions. These restricted and non-restricted

Fig. 7.3 Semantic segmentation model output

Fig. 7.4 Floor space mapping

regions vary depending on the use case. We have demonstrated the same for our particular use case, efficient and aesthetic plant placement.

Restricted or Red zones are defined as those regions where no plants should be placed so that they do not block any pathway like in front of the door or where the furniture lays. To mark the restricted region (red zone), a proportion of empty space area in front of the door is identified and marked as red as shown in the Fig. 7.5. Non-Restricted or green zones are defined as those regions where we can place our plants, like in front of railing, walls etc. Using this, a proportion of non-restricted

Fig. 7.5 Red and green zone identification

empty space is identified in front of railings and walls and marked as green as shown in Fig. 7.5.

7.3.6 Area Calculation and Utilisation for Plant Placement

The custom algorithm mentioned in Sect. 7.3.4 determines the total empty floor space area. The logic presented in Sect. 7.3.5 helps in the marking of red and green zones. In our algorithm, initially the area for all green zones is computed using built-in OpenCV function.

Any line can be represented as c = ax + by. Considering two points P and Q satisfy the given line equation, we can set the values so that this equation holds true. For identification of points in green zone shown in Figure 7.5 where plants should be placed, a line equation, 1 is written using the two vertices (P and Q) of identified rectangle as mentioned below:

$$\begin{aligned} a &= Q[y] - P[y] \\ b &= P[x] - Q[x] \\ c &= a * (P[x]) + b * (P[y]) \end{aligned} \tag{7.1}$$

where, P(x,y) and Q(x,y) are the two endpoints of the railing and a and b are variables of linear Eq. 7.1.

Fig. 7.6 Bounding box creation in the Green Zone

If b is less than 0, Eq. 7.2 represents the line passing through points P and Q else, Eq. 7.3 represents the same.

$$ax - by = c \quad (7.2)$$

$$ax + by = c \quad (7.3)$$

Using the above line equation, we get the bounding box coordinates for the first plant placement. To place the next bounding box we have kept spacing of 200 pixels. The space between two adjacent boxes is not considered while placing plants so that none of the images overlap. These steps are iterated till the bounding box coordinates exceed the green zone and thus, the total number of plants that can be placed in one row is calculated, refer Fig. 7.6.

7.3.7 Image Blending

Here, we are using the mapped empty floor space for placing plants and generating balcony garden design recommendation. We have a repository of plant images from which our algorithm randomly selects the foreground images to put in the identified empty regions. We can add the plant images directly to the empty regions using the coordinates from semantic segmentation model but then both the images would not be seamlessly blended as they have some background associated to them. To make the final image more authentic and real, foreground plant images are randomly picked and merged with the background using our image blending module purely based on OpenCV [14].

Blending images of varying sizes involves three major stages: selecting, masking and overlaying. To blend images of different sizes, the module first selects a Region Of Interest (RoI) in the background image which in our case is the identified empty region. The next step is to replace the values in that particular region with the values of the foreground image and then add them back in. However, here we plan to transfer a specific region and do not want the background from the foreground image. To achieve this we performed multiple masking and bit-wise operations.

The foreground image is first resized to fit the co-ordinates of the identified regions (RoI) detected in the background image. The resized image is gray scaled and thresholded to create a binary image (mask). In order to bring back the original colors, we apply image processing to inverse the mask using bit-wise operations. The final blended portion is then generated by adding our foreground and RoI, which is then overlaid over the original background image.

7.4 Result and Conclusion

Our approach offers an end to end solution for floor mapping and empty floor space detection which is a key requirement across many domains and multiple use cases. There are already implemented approaches available for floor mapping in the 3D space but there's no such implementation for 2D space. Because of the third dimension i.e. depth, present in 3D, the remoteness of objects in images is easily determined. This aids in accurate identification and segmentation of objects in the image in 3D space.

However, identifying objects in a 2D space is a bit challenging due to the lack of depth parameter. Here, we only have two parameters to work around, and yet we are able to achieve strikingly accurate results. Using the segmentation approach, we are able to implement empty floor space identification efficiently for plant placement, shown in Fig. 7.7.

Test images are manually annotated for six classes (wall, door, window, ceiling, railing and furniture) to yield ground truth masks. The predicted masks we get from our model are compared with the ground truth. The performance metrics that we have used here is Intersection Over Union (IoU) which is one of the most efficient accuracy metrics applied for semantic segmentation based use cases. IoU is defined as the area of overlap between the predicted segmentation and the ground truth divided by the area of union between the predicted segmentation and the ground truth, as shown in Fig. 7.8.

This metric ranges from 0–1 (0–100%) with 0 signifying no overlap and 1 signifying perfectly overlapping segmentation. The IoU that we are able to achieve on our dataset is 0.964 (96.4%) for a threshold of 0.5.

The train, test and validation accuracies achieved for our use case are as follows -

Fig. 7.7 Output image with 1 or 2 row of plant

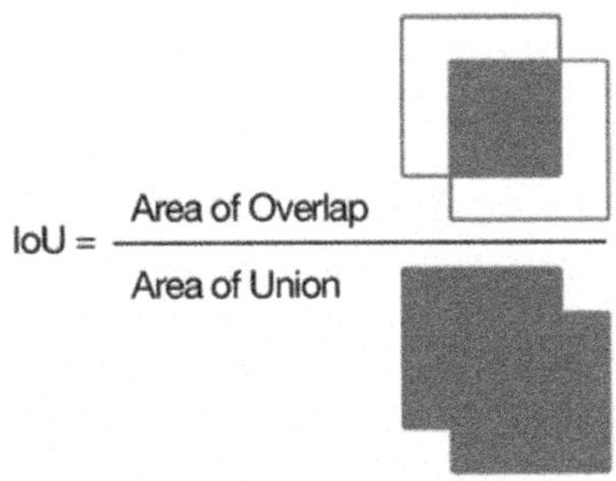

Fig. 7.8 IoU calculation visualized. *Source* Wikipedia

Train Accuracy – 98%
Test Accuracy – 96%
Validation Accuracy – 95%

From the above results, we can see that our approach is capable of effectively identifying empty floor spaces and classifying them into restricted and non-restricted areas even when the images are taken from different camera angles.

For our future work, we plan to extend the proposed technique to other use cases like indoor design recommendations, autonomous robot navigation, empty parking space identification, and compare them with other state-of-the-art methods.

References

1. Wang, P., Chen, P., Yuan, Y., Liu, D., Huang, Z., Hou, X., Cottrell, G.: Understanding convolution for semantic segmentation. In: 2018 IEEE Winter Conference on Applications of Computer Vision (WACV), pp. 1451–1460 (2018)
2. Chen, L., Zhu, Y., Papandreou, G., Schroff, F., Adam, H.: Encoder-decoder with atrous separable convolution for semantic image segmentation. CoRR, abs/1802.02611 (2018). http://arxiv.org/abs/1802.02611
3. Chen, L., Papandreou, G., Schroff, F., Adam, H.: Rethinking atrous convolution for semantic image segmentation. CoRR, abs/1706.05587 (2017). http://arxiv.org/abs/1706.05587
4. Girshick, R.: Fast r-cnn. In: Proceedings of the IEEE International Conference on Computer Vision (ICCV), December (2015)
5. Ren, S., He, K., Girshick, R., Sun, J.: Faster r-cnn: towards real-time object detection with region proposal networks (2016)
6. He, K., Gkioxari, G., Dolla´r, P., Girshick, R.B.: Mask R-CNN. CoRR, abs/1703.06870 (2017). http://arxiv.org/abs/1703.06870
7. Abdulla, W.: Mask r-cnn for object detection and instance segmentation on keras and tensorflow (2017). https://github.com/matterport/MaskRCNN
8. Lin, T., Maire, M., Belongie, S.J., Bourdev, L.D., Girshick, R.B., Hays, J., Perona, P., Ramanan, D., Dolla´r, P., Zitnick, C.L.: Microsoft COCO: common objects in context. CoRR, abs/1405.0312 (2014). http://arxiv.org/abs/1405.0312
9. R. I. of Technology: Floor plan from rakuten real estate and pixelwise wall label. https://rit.rakuten.co.jp/datarelease/. 23 April (2019)
10. Liu, C., Wu, J., Kohli, P., Furukawa, Y.: Raster-to-vector: Revisiting floorplan transformation. In: Proceedings of the IEEE International Conference on Computer Vision (ICCV), October (2017)
11. Kalervo, A., Ylioinas, J., Ha¨ikio¨, M., Karhu, A., Kannala, J.: Cubicasa5k: a dataset and an improved multi-task model for floorplan image analysis. CoRR, abs/1904.01920 (2019). http://arxiv.org/abs/1904.01920
12. Liu, C., Wu, J., Furukawa, Y.: Floornet: A unified framework for floorplan reconstruction from 3d scans. CoRR, abs/1804.00090 (2018). http://arxiv.org/abs/1804.00090
13. He, K., Gkioxari, G., Dollar, P., Girshick, R.: Mask r-cnn. In: Proceedings of the IEEE International Conference on Computer Vision (ICCV), October (2017)
14. Culjak, I., Abram, D., Pribanic, T., Dzapo, H., Cifrek, M.: A brief introduction to opencv. In: 2012 Proceedings of the 35th International Convention MIPRO, pp. 1725–1730 (2012)

Chapter 8
Light-Field Camera Based Spatial Multiple Small Particle Tracking with Post-processing

Miao Zhang, Jörg Matthes, Krasimir Aleksandrov, Hans-Joachim Gehrmann, and Markus Vogelbacher

Abstract Multi-object detection and tracking is one of the most significant tasks in computer vision, which has been widely applied in various fields, such as biomedicine, automatic driving, and video surveillance. In our project, we aim to investigate the flight behavior of different small refuse-derived fuel particles (e.g., plastic, wood, paper, etc.) with sizes around 5 to 30 mm, which are blown into a combustion chamber (rotary kiln) via a lance, by detecting and tracking the particles. By utilizing a calibrated light-field camera, both 2D gray value images and 3D information, including the depth of the scene, are available for investigation. To obtain particle trajectories, accurate temporal localization of the particles is necessary. Thus, we perform an approach combining a 2D detection method and a 3D clustering method for precise detection of the particles at first. Subsequently, the temporal 2D particle localizations are associated into tracklets based on the Kalman filter state estimator and the global nearest neighbor algorithm. Owing to the wrong and incomplete tracklets caused by several inaccurate detections, we present a post-processing method to enhance the tracking performance in the paper. Finally, the optimized 2D particle trajectories are converted into 3D trajectories, and these obtained 3D particle trajectories are analyzed by a polynomial regression for further analysis of the particle flight behavior.

Keywords Multiple object tracking · Tracklet connection · 3D particle trajectories · Light-field camera

M. Zhang (✉) · J. Matthes · M. Vogelbacher
Institute for Automation and Applied Informatics, Karlsruhe Institute of Technology, Karlsruhe, Germany
e-mail: miao.zhang@kit.edu

K. Aleksandrov · H.-J. Gehrmann
Institute for Technical Chemistry, Karlsruhe Institute of Technology, Karlsruhe, Germany

M. A. Laribi et al. (eds.), *Advances in Automation, Mechanical and Design Engineering*, Mechanisms and Machine Science 121,
https://doi.org/10.1007/978-3-031-09909-0_8

8.1 Introduction

With the development of advanced computer vision technology and corresponding algorithms, multiple object detection and tracking becomes one of the most crucial and challenging tasks with a wide range of applications, such as particle tracking velocimetry (PTV) for flow analysis, pedestrian surveillance, and biomedicine. Tracking-by-detection is a dominant approach for multiple-object tracking (MOT), which at first detects objects in each video frame and associates the detections to several tracks subsequently. In our work, we utilize this approach to investigate the flight behavior of various small refuse-derived fuel particles moving in a rotary kiln by analyzing obtained spatial particle trajectories based on images captured by a light-field camera. The acquired knowledge concerning the flight behaviors, e.g., flight trajectories and velocities, is conducive to a controllable and secure application of the fuel particles. Moreover, CFD models related to fuel combustion are developed simultaneously under the same project framework, and the experimental results are capable of validating the models.

8.1.1 Experiment Setup

Via a lance, the fuel particles are blown into the rotary kiln, which has a length of 8.4 m and an inner diameter of 1.4 m. The kiln is part of the BRENDA facility at the Institute of Technical Chemistry, Karlsruhe Institute of Technology. Since the utilization of a stereo camera system is impossible due to constructional constraints of the rotary kiln, a light-field camera is mounted outside the kiln to record the inner 3D scenes through a quartz glass window at the outlet of the kiln. The light-filed camera R12 produced by the company Raytrix with a lateral resolution of 3072×4096 pixels and a framerate of 330 fps allows the conversion of 2D coordinates of particles into 3D with depth information. Therefore, both 2D image gray values and 3D point coordinates are available for further image processing, as illustrated in Figs. 8.1 and 8.2. Figure 8.1b is known as a depth map that depicts the calculated depth information as a false-color image.

Nevertheless, due to the light-field camera setup, there are significant fluctuations in the depth measurement of the light-field camera, which have to be considered in the subsequent processing, i.e., particle detection and tracking.

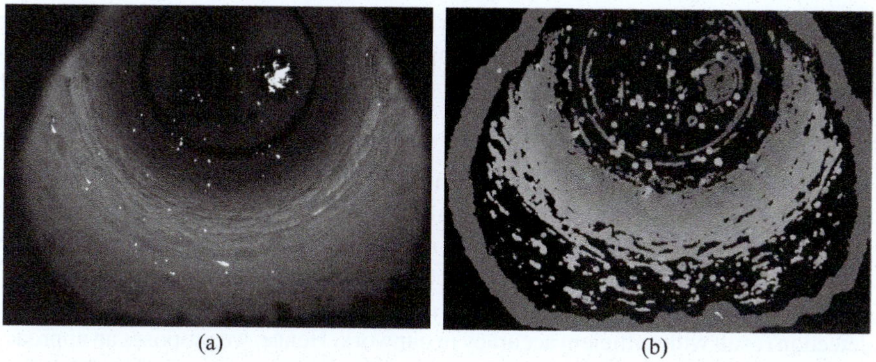

Fig. 8.1 Examples of diverse images. **a** Basic-focus image captured by light-field camera, **b** Corresponding depth map in false color

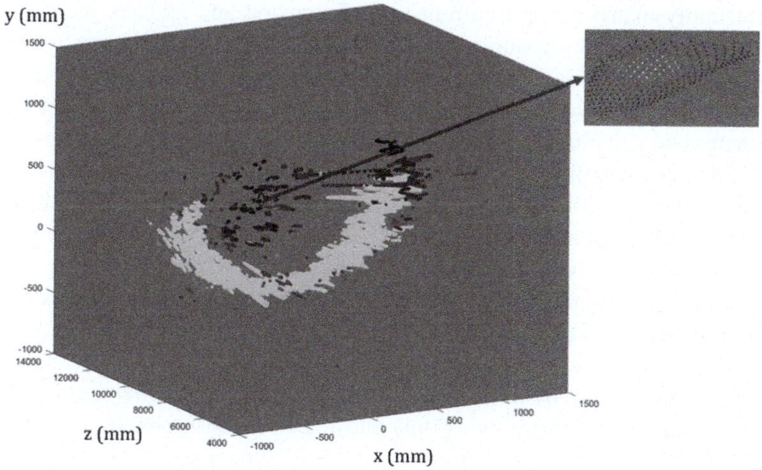

Fig. 8.2 Example of a 3D point cloud

8.1.2 Related Work

As mentioned above, the analysis of the flight behaviors of various refuse-derived fuels can be realized by tracking-by-detection, where object detection and data association are the key tasks. Over the last several decades, many researchers contribute to the development of efficacious algorithms for object detection and tracking.

Plenty of detection approaches base on the gray value of the captured image, such as the OTSU threshold selection method that selects a threshold automatically from a gray-level histogram for extracting foreground objects [1] and the method Scale Invariant Feature Transform (SIFT), which identifies objects by extracting highly distinctive features in various scales of the image [2, 3]. With the development of machine learning, artificial neural networks have also been taken into consideration for object detection [4, 5]. For spatial object detection, clustering methods such as mean shift clustering [6] and density-based spatial clustering of applications with noise (DBSCAN) [7] are implementable when image information in the form of a 3D point cloud is available.

Since the temporary locations of each particle are the most crucial input of the tracking process, the particle detection results strongly affect the performance of multi-object tracking. False detections, missing detections, or duplications show a considerably negative impact on the accuracy of the tracking results. Neither 2D detection method based on gray value nor clustering method is capable of providing a detection result with sufficient accuracy in our work. Hence, we proposes an approach that combines the 2D detection method SIFT and the 3D Clustering method DBSCAN for precise spatial particle detection [8].

Generally, a multiple object tracking approach consists of two crucial parts: state estimation and data association. The state estimator predicts the state vector and its uncertainty matrix for the current time dimension and updates them afterward based on the assigned measurement by a data association algorithm, which associates detections with tracks according to a predefined cost function. The state vector depends highly on the scope of application and normally consists of temporary object coordinates and velocities for object tracking. A widely applied state estimator is the linear Kalman filter [9] that is recursive and optimal under the assumption that the errors are normally distributed. Further extensions and generalizations of the algorithm, such as the extended Kalman filter [10] and the unscented Kalman filter [11], are developed to deal with nonlinear systems.

One commonly used approach for data association is the global nearest neighbor (GNN) approach, where the cost function matrix is composed of the Mahalanobis distance between the detections and the currently predicted state vectors of the tracks. The assignment follows by finding the global minimum cost utilizing the Hungarian algorithm [12]. Instead of assigning only one detected measurement to a certain track and discarding the others, Bar-Shalom and Tse [13] used probabilistic data association (PDA) that weights all measurements within the gating region of the track to follow targets in a cluttered environment. Since PDA considers all possible measurements, the negative influence caused by clutter leading to wrong tracking results can be reduced. Nevertheless, PDA assumes that tracks are isolated from each other, i.e., the situation that different tracks share the same measurements has not been paid much attention. Therefore, Fortmann et al. [14] proposed the joint probabilistic data association (JPDA) algorithm that computes joint posterior association probabilities for multiple object tracking. As an improvement of JPDA, Mušicki and Evans [15] proposed an algorithm known as joint integrated probabilistic data association (JIPDA), which utilizes the probability of object existence and results in recursive expressions for the probability and data association coefficients. Reid [16] demonstrated a sophisticated algorithm known as multiple hypothesis tracking (MHT) that solves the tracking problem by evaluating the likelihood of each possible tracking hypothesis.

Despite the efforts to optimize detection approaches, the mentioned issues remain inevitable due to characteristics of employed measurement equipment, illumination problems because of the reflection behavior of the particles, and the applied detection methods. These detection problems can lead to separating one complete particle trajectory into several tracklets and thereby influence the tracking performance. Some researchers proposed developing more precise and robust tracking filters or using more sophisticated methods for data association to enhance the tracking performance, such as JPDA and MHT. Nevertheless, these methods are not adequate in some complex occasions, such as in our case, when many small particles without proper representative features are blown from a fixed lance within a short time interval, as shown in Fig. 8.3.

To obtain accurate and complete 2D particle trajectories, we present a post-processing method to combine and merge tracklets into complete trajectories of the particles. Some researchers demonstrated tracklets reconnection approaches based on the appearance similarity of the representative objects and the motion similarity of the tracklets [17, 18]. However, these methods rely strongly on the representative features of the tracked objects, and hence, are primarily used for people tracking, which is not definitively appropriate for tracking small objects without any specific properties.

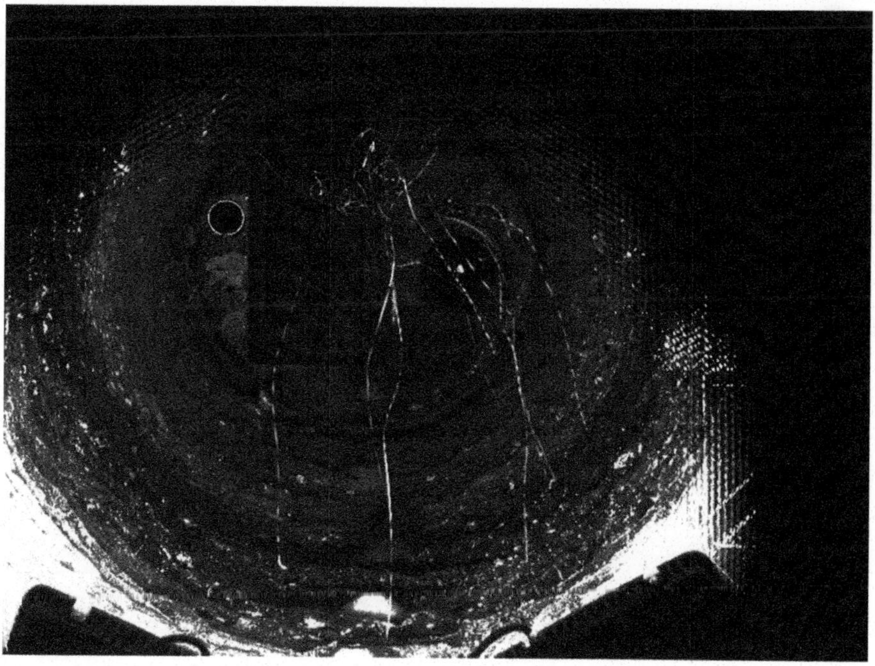

Fig. 8.3 Example of particle trajectories obtained by adding the images from a sequence

Under these circumstances, we present an extended tracking algorithm using 3D information for improving 2D tracking, and subsequently, we propose a post-processing method to connect and merge tracklets into complete trajectories without utilizing additional object features.

8.1.3 Work Overview

The flow chart in Fig. 8.4 schematically shows the overview of the entire work, viz. multiple particle tracking in the rotary kiln, on the basis of which the paper is organized. In Sect. 8.2, we elaborate on a new detection approach that identifies particles utilizing both 2D and 3D image information. The approach is proved to show a better performance than implementing a single detection method. Section 8.3 demonstrates the algorithm for MOT. A direct 3D tracking is currently not applicable in our work owing to the uncertain and, therefore, strongly fluctuating depth information (z-direction). Therefore, we track the particles in 2D at first and extract the 3D coordinates afterward. Besides, we also propose an approach for post-processing, including tracklet elimination, tracklet connection/merging, and perhaps trajectory fusion. In Sect. 8.4, the obtained 3D trajectories are estimated with polynomials

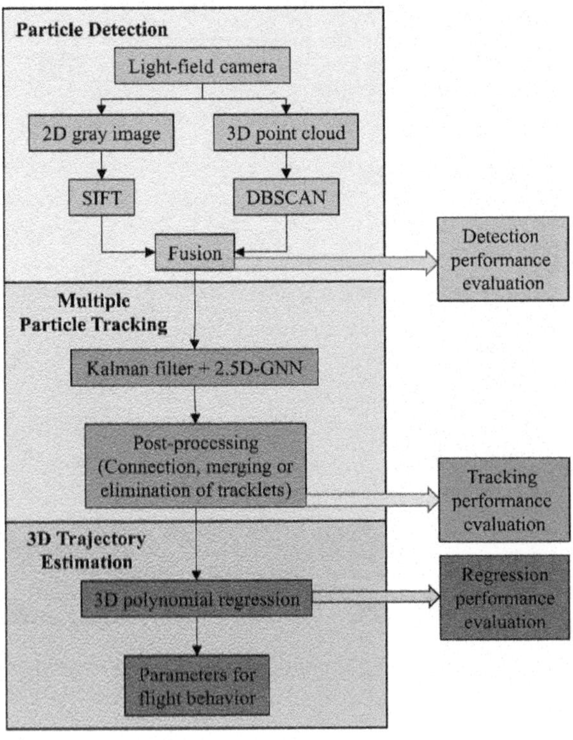

Fig. 8.4 Flow chart of the entire work

based on a regression approach. Section 8.5 provides the experimental validation of the accuracy of the proposed approach. For evaluation of multiple particle tracking, 2D locations of the captured particles are manually associated in accordance with their time dimensions into particle trajectories, which serve as the ground truth. The tracking performance is evaluated by exploring an evaluation software recommended in [19]. Section 8.6 concludes the paper.

8.2 Particle Detection

In our project, both 2D image gray values and 3D point clouds are obtainable owing to the application of the light-field camera. Thus, we present a combination method that combines detection results of 2D SIFT and 3D DBSCAN to detect multiple small particles accurately. Firstly, the two approaches are executed separately, and the results are compared with each other afterward. The comparison occurs two-dimensionally, whereby the result obtained by the clustering method will be transformed into 2D at first. If the particle positions and radii are taken into account, the following cases are possible during the comparison:

1. A potential particle is detected as precisely one particle by both methods.
2. A potential particle is detected as one particle by one method but is detected as more than one particle by another.
3. A potential particle is only detected by one method.

For case 1, the particle candidate will be directly accepted in the final detection result. For the other two cases, when the results of the methods are not identical, gray values of the particle region will be analyzed to determine the accurate number and position of particles.

The following reasons are responsible for the different detection results of the same particle region:

- Some particles are too small and dark to be detected by gray value based object detection approaches.
- A few particles are spatially so close to each other that the clustering method is not capable of segmentation the point cloud.
- For a big particle, especially with strongly inhomogeneous gray value distribution, the gray value based detection method might wrongly identify the particle as several dispersed particles.

By detailed analysis of the gray value in the divergent particle region, the final detection result containing both 2D and 3D particle information is obtained. Since the combination approach takes advantage of both 2D (gray value) and 3D (point cloud) image information, it shows a more accurate performance than only executing one method, which has been proved in [8], e.g., while 83.3% of the particles are detected by the DBSCAN clustering method and 84.4% by the SIFT approach, the combined method achieves a ratio of correct particle identification of 90.0%.

8.3 Multiple Particle Tracking

In this section, the complete tracking process, including post-processing, is elaborated.

8.3.1 2.5 D Multiple Particle Tracking

Because of the application of a light-field camera with up to 330 fps, the particle movement within two consecutive frames can be approximated as linear. Due to the enhanced accuracy of the combination method illustrated in Sect. 8.2, the total amount of false detections and duplications are reduced. Hence, we use the linear Kalman filter and the GNN approach for multiple particle tracking currently.

Unlike standard 2D object tracking with GNN utilizing simply the Mahalanobis distance for gating, we use Mahalanobis distance together with movement in depth instead. For the assignment of a unique track, not only the measurements with a Mahalanobis distance to the current predicted position of the track larger than a defined threshold value will be discarded, but also the measurements with a movement in depth exceeding a limit will be removed from the potential assignment list.

The cost matrix for data association is generated afterward. Assuming a current data association with n tracks and m measurements, [12] defines the cost matrix as a n × m matrix with the following element values:

$$c_{ij} = \begin{cases} C, if\ measurement\ j\ is\ not\ in\ the\ gate\ of\ track\ i \\ d_{ij}^2 + ln|S_i|, if\ measurement\ j\ is\ in\ the\ gate\ of\ track\ i \end{cases} \quad (8.1)$$

where C denotes a predefined cost for non-assignment; d_{ij}^2 represents the 2D Mahalanobis distance between track i and measurement j;S_i is the residual covariance matrix for track i at the current time dimension. $\ln|S_i|$ is a penalizing term for tracks with highly uncertain predictions, such as the tracks without assigned measurements for a period of time. For those tracks, S_i could be very large, which lead to smaller Mahalanobis distance and further to incorrect data association. Under this circumstance, the penalizing term is conducive to reducing the possibility of wrong assignments.

Since we employ depth information for particle tracking, we define our tracking as 2.5D. In principle, the cost function can consist of both Mahalanobis distance and motion in depth as well. Whereas owing to the strongly fluctuated depth information, the disadvantages of a quantitative application of depth coordinate in the cost function outweigh the advantages.

8.3.2 Post-Processing

As explained in Sect. 8.1, post-processing is necessary when dealing with complex tracking issues, especially tracking-by-detection with false detections, missing detections, or duplications. To facilitate tracking accuracy, we propose and demonstrate a framework illustrated in the flow chart below (Fig. 8.5). In our work, we consider all particle traces before tracklets connection as tracklets regardless of their actual lengths and durations. Moreover, all positions here refer to the detected coordinates of the particles. Since the ground truth is generated optically on the basis of captured images, utilizing the detected positions instead of the updated coordinates by the Kalman filter benefits an objective and accurate assessment of the tracking performance. Besides, all notations with subscript ϵ indicate threshold values that are elaborated in Sect. 8.5.

Faulty Tracklets Elimination. First of all, we eliminate faulty tracklets that are too short, too slow, or too irregular. A track T_i can only persist when it satisfied the following three conditions:

$$\mathcal{D}(T_i) > \mathcal{D}_\epsilon \tag{8.2}$$

$\mathcal{D}(T_i)$ denotes the duration of T_i that counts the number of particle coordinates in T_i.

$$\left|\overline{V_i}\right| > \left|\overline{V}\right|_\epsilon \tag{8.3}$$

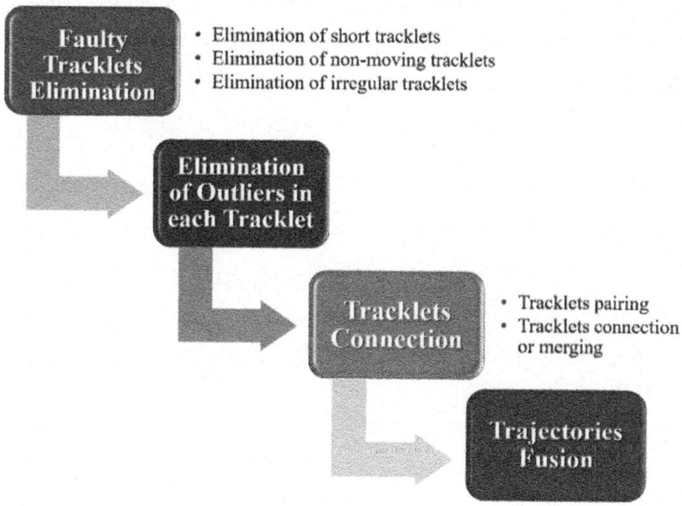

Fig. 8.5 Flow chart of the post-processing framework

$\left|\overline{\boldsymbol{V}_i}\right|$ stands for the average magnitude of the velocity vector of T_i, which is formulated as:

$$\left|\overline{\boldsymbol{V}_i}\right| = \frac{\sum_{t=t_1}^{t=t_{end-1}} \left|\boldsymbol{V}_{i,t}\right|}{\mathcal{D}(T_i) - 1}, \tag{8.4}$$

with

$$\boldsymbol{V}_{i,t} = \frac{\boldsymbol{T}_{i,t+\Delta t} - \boldsymbol{T}_{i,t}}{\Delta t}, \tag{8.5}$$

is the temporary velocity of T_i at time t that is calculated as the difference of two consecutive positions $\boldsymbol{T}_{i,t+\Delta t}$ and $\boldsymbol{T}_{i,t}$ dividing by their time difference Δt.

$$R_{T_i \tilde{T}_i} > R_\epsilon \tag{8.6}$$

In terms of the ground truth, the 2D particle trajectories roughly correspond with second-degree polynomials. We, therefore, approximate T_i by a second-degree polynomial $\tilde{T}_i$ and $R_{T_i \tilde{T}_i}$ stands for the cross-correlation coefficient between T_i and $\tilde{T}_i$.

Only the tracklets with the abovementioned three measures exceeding certain threshold values remain for further processing.

Elimination of Outliers in Each Tracklet. After the elimination of the faulty tracklets, the outliers in each tracklet will be subsequently deleted. This step is considered a preparation for tracklet connection and is conducive to finding potential tracklet connection pairs. A particle position at time t belonging to track T_i is regarded as outlier, if

$$\left|\Delta \boldsymbol{V}_{i,t}\right| > \left|\Delta \boldsymbol{V}\right|_\epsilon \,\&\, \theta_{i,t} > \theta_\epsilon, \tag{8.7}$$

with

$$\left|\Delta \boldsymbol{V}_{i,t}\right| = \left|\boldsymbol{V}_{i,t} - \boldsymbol{V}_{i,t-1}\right|, \tag{8.8}$$

and $\theta_{i,t}$ denoting the angle between $\boldsymbol{V}_{i,t}$ and $\boldsymbol{V}_{i,t-1}$. Figure 8.6 depicts schematically $\boldsymbol{T}_{i,t}$, $\boldsymbol{V}_{i,t}$, and $\theta_{i,t}$.

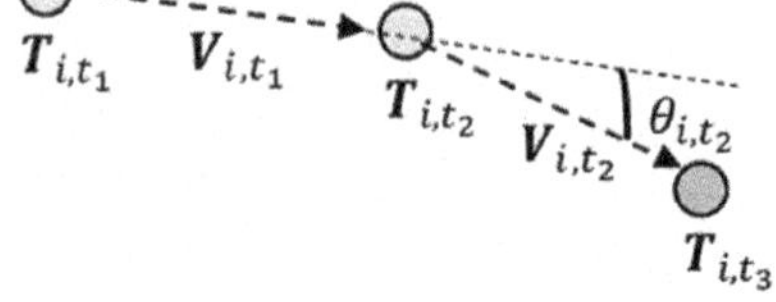

Fig. 8.6 Schematic illustration of $T_{i,t}$, $V_{i,t}$, and $\theta_{i,t}$

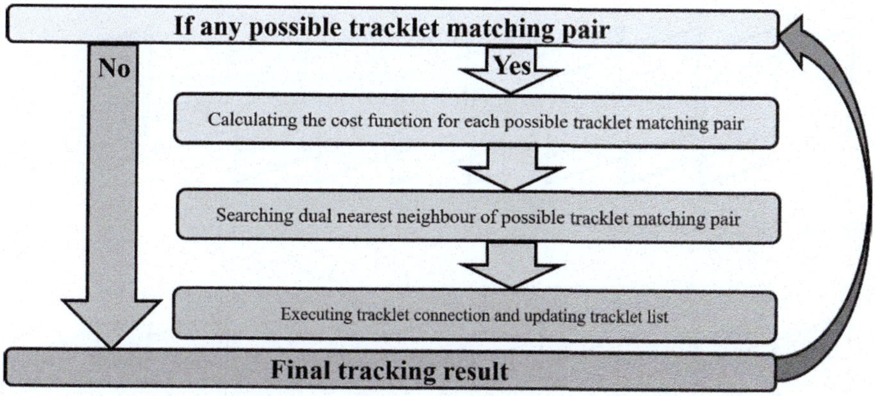

Fig. 8.7 Procedure of tracklet connection

Tracklet Connection

Since the faulty tracklets and the outliers in the tracklets are deleted, we are able to associate the remaining tracklets into trajectories. Figure 8.7 illustrates the connection process, which starts with scanning for possible tracklet matching pairs. Subsequently, the cost function of each possible matching pair is calculated. Only the bilateral nearest neighbor matching pairs are allowed for connection, and afterward, the entire tracklets list is updated. This procedure will be iterated until there is no possible tracklet matching pair. Criteria for a possible tracklet matching pair and the calculation of the corresponding cost function are elaborated in the following content.

(1) Criteria for possible tracklet matching pair

Assuming two tracklets T_i and T_k with T_i ending earlier than T_k, we can probably expect two situations concerning their appearance: 1. T_i and T_k do not overlap in the timeline (case 1), 2. part of T_i and T_k appear at the same time (case 2). For case 1, the first matching criterion is determined as follows:

$$t_{T_k,1} - t_{T_i,end} < t_{\epsilon,1}, \tag{8.9}$$

where $t_{T_i,end}$ denotes the end time of T_i and $t_{T_k,1}$ denotes the first appearance time of T_k. Only if the time gap of the two tracklets is shorter than a threshold value $t_{\epsilon,1}$, these two tracklets can be considered as a matching pair and will possibly be simply connected, as shown in Fig. 8.8.

If two tracklets overlap in timeline, the first matching criterion is converted to

$$t_{T_i} \cap t_{T_k} < t_{\epsilon,2}, \tag{8.10}$$

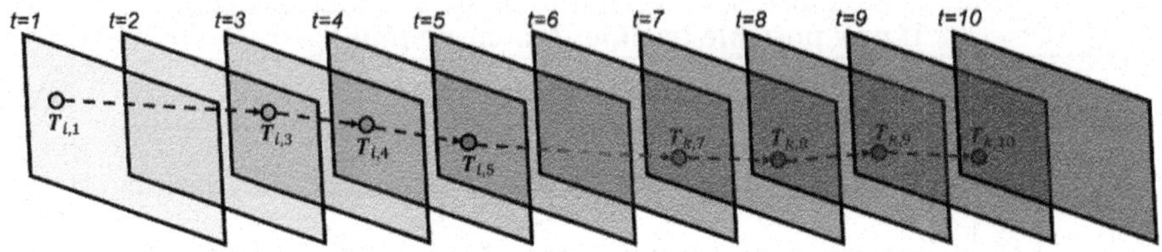

Fig. 8.8 Tracklet connection

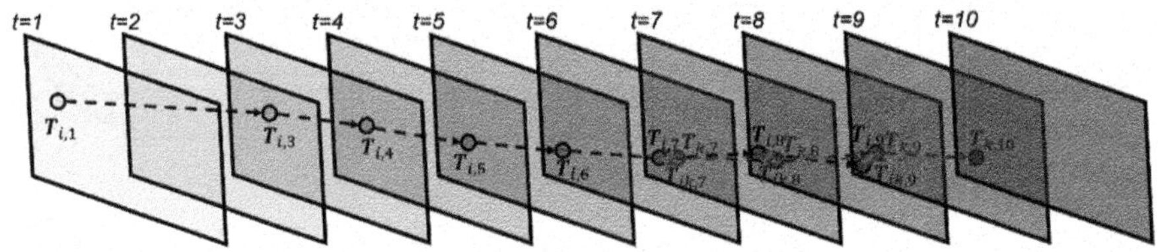

Fig. 8.9 Tracklet connection and merging

which indicates a time overlap of T_i and T_k shorter than a threshold value $t_{\epsilon,2}$. Since part of the matching pair appear at the same time, the tracklets are not only connected but are merged into one at the overlapped period, as illustrated in Fig. 8.9.

Obviously, a simple application of the time constraint is not sufficient for scanning potential tracklet matching pairs since the locations and tendencies of the tracklets can vary widely. Therefore, we define further criteria with respect to characteristics of the newly generated tracklet in the connected domain T_{ik}, which connects the three last temporary locations of T_i (e.g., Fig. 8.10, left, $t = 5,7$, and 8) and three first locations of T_k (e.g., Fig. 8.10, left, $t = 11,13$, and 14) for case 1. For case 2, T_{ik} consists of the first temporally coincident location (e.g., Fig. 8.10, right, $t = 9$) together with its two previous locations of T_i (e.g., Fig. 8.10, right, $t = 5$ and 7), the last temporally coincident location (e.g., Fig. 8.10, right, $t = 11$) together with its two following locations of T_k (e.g., Fig. 8.10, right, $t = 13$ and 14), and the merged tracklet in between. The merged tracklet takes the midpoint of the particle positions from the two tracklets for overlapped time dimension (e.g., Fig. 8.10, right, $t = 9$ and 11) and simply takes the particle positions from one tracklet for no overlapped time.

Only if T_{ik} satisfies the three requirements below, the two tracklets T_i and T_k can be regarded as potential matching pair, whose matching cost will be calculated later.

- Requirement 1: No outliers exists in connected/merged tracklet T_{ik} according to subsection 3.2.2. As the outliers in the original tracklets are already eliminated, any outlier in T_{ik} can indicate an inadequate connection of T_i and T_k.
- Requirement 2: $R_{T_{ki}\tilde{T}_{ki}} > R_\epsilon$.
- Requirement 3: A strong fluctuation of T_{ik} in the depth direction is not allowed, viz.

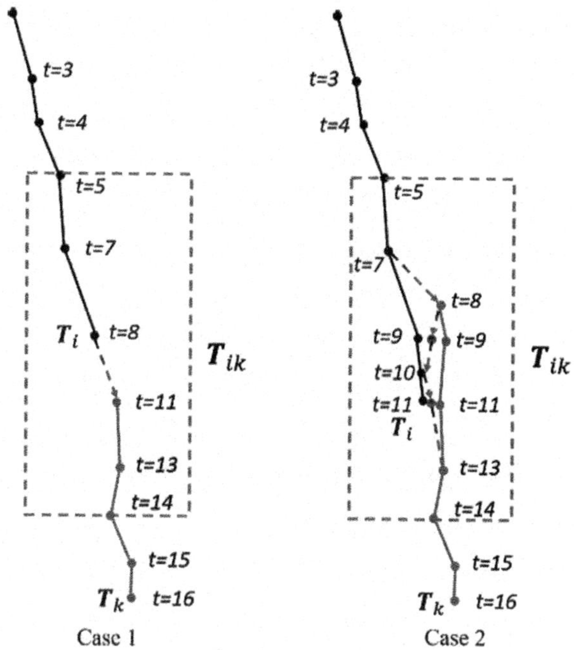

Fig. 8.10 Schematics of tracklet connection and tracklet merging

$$\Delta z_{ik,t} < z_\epsilon \tag{8.11}$$

with

$$\Delta z_{ik,t} = \left| \frac{z_{T_{ik,t+\Delta t}} - z_{T_{ik,t}}}{\Delta t} \right|. \tag{8.12}$$

(2) Computation of the cost function

As mentioned above, the cost of each possible matching pair is computed subsequently. The cost function takes not only the previously discussed characteristics of the generated tracklet T_{ik} but also the gray value in the connected domain into consideration. The function is formulated as:

$$\begin{aligned} C(T_i, T_k) = a\bar{\theta}_{ik} + b\left|\overline{\Delta \boldsymbol{V}}_{ik}\right| + c\overline{\Delta z}_{ik} + d \cdot \max((t_{T_i,end} - t_{T_k,1}), 0) \\ + e\left(t_{T_i} \cap t_{T_k}\right) + \frac{f}{\max\left(\bar{I}_{ik}, 1\right)}, \end{aligned} \tag{8.13}$$

where a, b, c, d, e and f are weighting factors for each term that should be determined in accordance with the magnitudes of the terms. The first three terms calculate respectively the arithmetic average of all $\theta_{ik,t}$ with the unit degree, the arithmetic average of all $\left|\Delta \boldsymbol{V}_{ik,t}\right|$ with the unit pixel per frame, and the arithmetic average of all $\Delta z_{ik,t}$ with the unit millimeter per frame in T_{ik}. The fourth and fifth terms represent the temporal cost of connection for case 1 and case 2. For case 1, the fifth term equals zero since the tracklets are not temporally overlapped. For case 2, the fourth term equals zero. The last term takes the gray value in the connected domain D_{ik} into account, where D_{ik} need schematizing at first. According to the movement in T_{ik}, the connected domain can be vertical or lateral. If the vertical motion is dominant, the domain is vertical, and vice versa. For vertical domain, D_{ik} is defined as a parallelogram with the following vertexes:

$$\left(\boldsymbol{T}^x_{i,t_{T_i,end}} - w, \boldsymbol{T}^y_{i,t_{T_i,end}}\right), \left(\boldsymbol{T}^x_{i,t_{T_i,end}} + w, \boldsymbol{T}^y_{i,t_{T_i,end}}\right),$$

$$\left(\boldsymbol{T}^x_{k,t_{T_k,1}} - w, \boldsymbol{T}^y_{k,t_{T_k,1}}\right), \left(\boldsymbol{T}^x_{k,t_{T_k,1}} + w, \boldsymbol{T}^y_{k,t_{T_k,1}}\right).$$

Accordingly, the 4 vertexes for a lateral domain D_{jf} are:

$$\left(\boldsymbol{T}^x_{j,t_{T_j,end}}, \boldsymbol{T}^y_{j,t_{T_j,end}} - h\right), \left(\boldsymbol{T}^x_{j,t_{T_j,end}}, \boldsymbol{T}^y_{j,t_{T_j,end}} + h\right),$$

$$\left(\boldsymbol{T}^x_{f,t_{T_f,1}}, \boldsymbol{T}^y_{f,t_{T_f,1}} - h\right), \left(\boldsymbol{T}^x_{f,t_{T_f,1}}, \boldsymbol{T}^y_{f,t_{T_f,1}} + h\right).$$

as illustrated in Fig. 8.11. The width (length) of the connected domain is controlled by the parameter w (h), which requires adjustment concerning the particle size. The mean gray value $\overline{I}_{ik}$ in the last term of the cost function computes the average gray value of D_{ik} based on the sum of difference images from $t_{T_{ik},1}$ to $t_{T_{ik},end}$, which are obtained by subtracting background from current images. In the light of prior knowledge with respect to particle sizes and statistical information regarding particle movements, we decide to set $a = 2$, $b = 2$, $c = 0.1$, $d = 3$, $e = 2$, $f = 600\,000$, and $w = h = 3$.

(3) Dual nearest neighbour

While the last two paragraphs address the problem of scanning possible matching pairs and computing their matching costs, the issue concerning finding an appropriate mean to connect the tracklets remains unsolved. Principally, the lower the cost of a matching pair is, the higher the probability of the connection will be. Therefore, utilizing the Kuhn–Munkres algorithm to find the global lowest cost (global nearest neighbor) might be a feasible solution. Nevertheless, it can lead to inaccurate connection results due to seeking a sole optimal solution, as depicted in Fig. 8.12, where green boxes mark the global lowest cost. C_{max} indicates the impossibility of tracklets connection and should be defined as larger than the maximal cost. While the fourth and fifth tracklets are correctly connected, the first and third tracklets are wrongly associated into one trajectory. Moreover, the second and fourth tracklets are also wrongly connected afterward. Under this circumstance, we recommend using dual nearest neighbor instead of the global nearest neighbor. Two tracklets are allowed

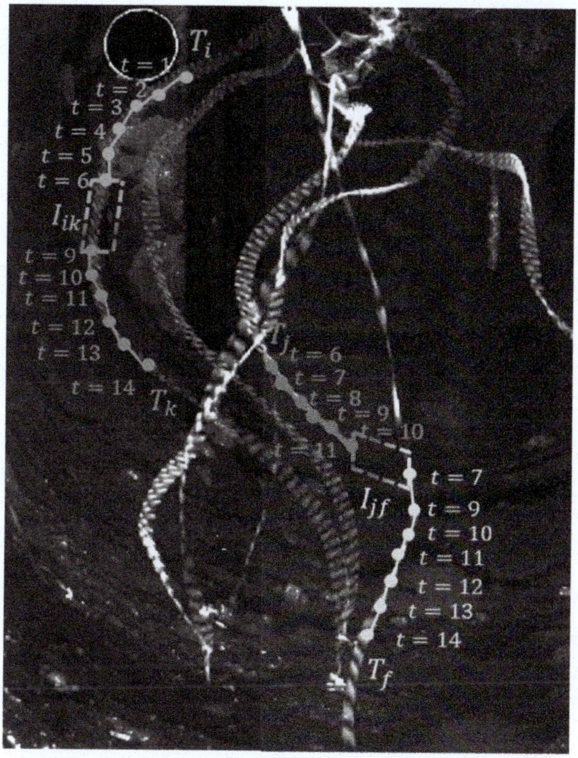

Fig. 8.11 Example of connected domain

to be connected only if they are the bilateral nearest neighbor, which means one tracklet is the nearest neighbor of the other, and meanwhile, the other one is also the nearest neighbor of the first. After the connection of all dual nearest neighbor pairs, the tracklet list will be updated, and the procedure will be repeated until there are no possible tracklet matching pairs. As schematically illustrated in Fig. 8.13, the tracklets are accurately connected into two trajectories after four iterations.

So far, we are able to acquire a solution for tracklet connection. The tracklets are then connected into trajectories according to the temporal overlap. For case 1, two tracklets are simply connected. Since several particles are sometimes invisible in the ground truth due to a combined effect of their rotations and illumination, the time gap between the tracklets is not filled with interpolated particle coordinates in order to obtain a relatively objective evaluation of the tracking performance later. For case 2, the newly generated tracklet in the connected domain is regarded as part of the trajectory at the corresponding time dimension and connected with the rest of the tracklets afterward.

As an alternative, we can compute the costs of all the tracklets matching pairs, e.g., if the total number of tracklets is n, the possible number of matching pairs amounts to $n(n-1)/2$, instead of seeking out possible tracklet matching pairs. Whereas a cost of non-assignment must be selected with high accuracy and adequacy to keep

Fig. 8.12 Example of tracklet connection using Kuhn–Munkres algorithm. (Tracklets with same marker belong to same trajectory.)

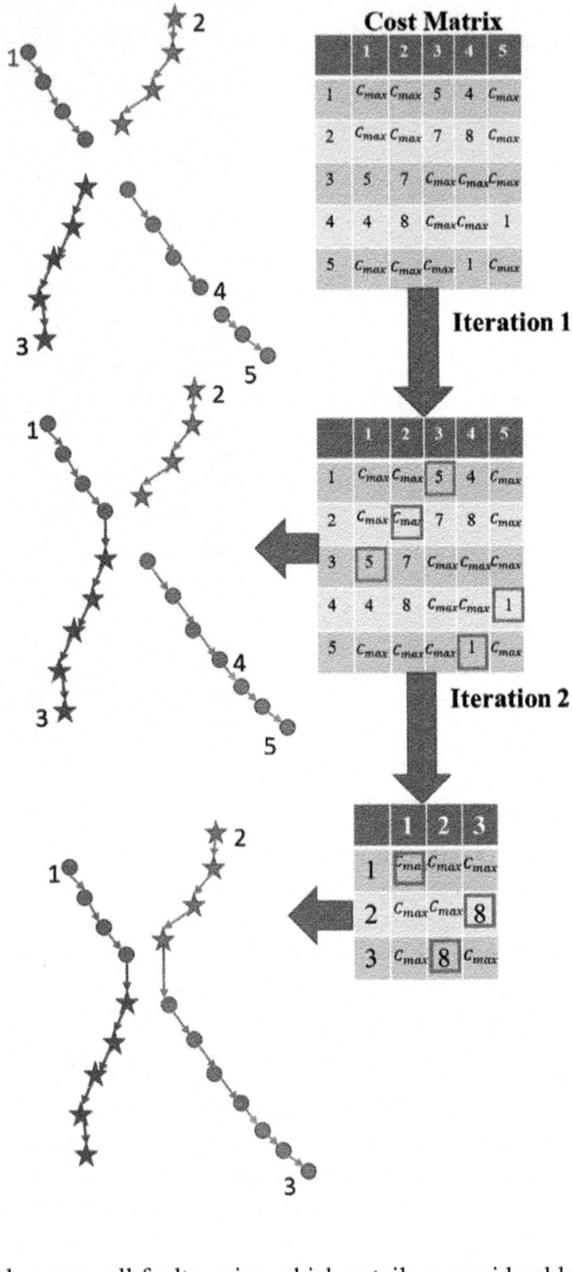

all correct matching pairs and remove all faulty pairs, which entails a considerable difficulty. Hence, we recommend scanning potential tracklet matching pairs at first for the sake of optimal connection performance.

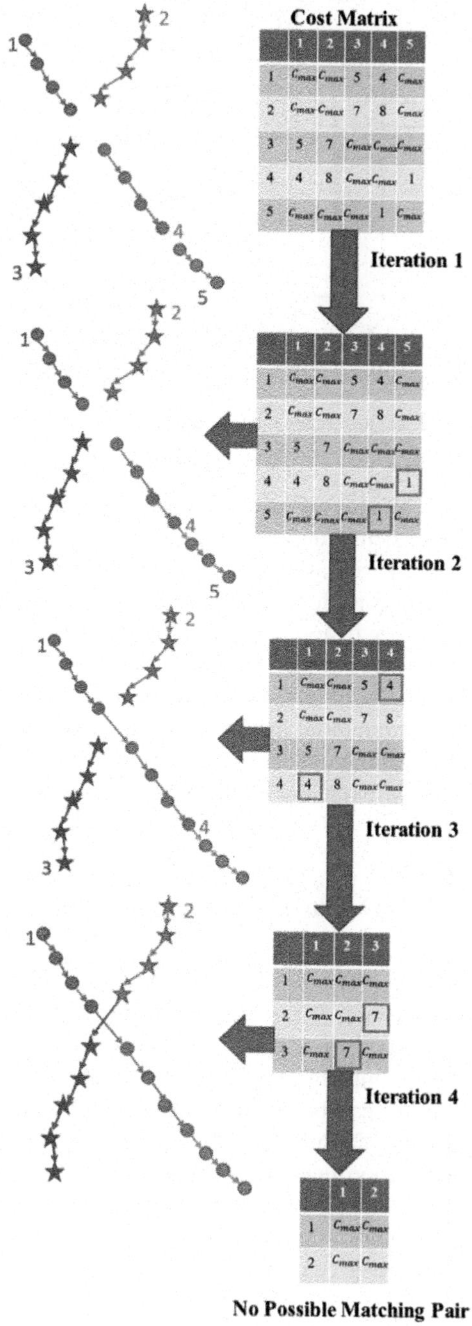

Fig. 8.13 Example of tracklet connection using dual nearest neighbor. (Tracklets with same marker belong to same trajectory.)

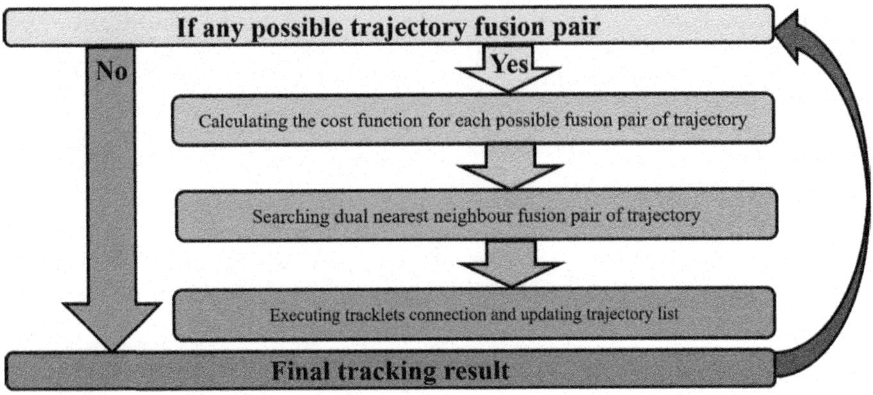

Fig. 8.14 Procedure of trajectory fusion

Trajectory Fusion. In case of a detection result with many duplications and wrong detections, which could cause one particle with more than one trajectory with a time period, the above proposed three procedures might not sufficient to acquire precise particle trajectories. Therefore, we recommend implementing a further step, trajectory fusion. The flow chart in Fig. 8.14 schematically presents the step.

Apparently, this process varies only slightly from the tracklet connection process. Possible tracklet matching pairs do not allow a long temporal overlap of two tracklets, but for potential fusion pairs, this constraint is ignored. Hence, the fourth and fifth terms in the cost function are pruned. Except for these distinctions, the rest of the procedure remains unchanged. An example of trajectory fusion is shown in Fig. 8.15.

8.4 3D Trajectory Estimation

Besides, 3D polynomial regression of the obtained particle trajectories via a fixed spatial point is implemented and evaluated based on parameters described in [20].

In order to obtain a mathematical description of the particle flight curves and thus to implement a detailed evaluation, the 3D particle trajectories with local fluctuations can be estimated with a polynomial in three independent dimensions. In each direction, the coordinate of a particle is a function of time, and three simultaneous coordinates determine the transient spatial location of the particle. Hence, the directional polynomial can be estimated separately and with different degrees. The following example shows a simple case where the motions are all estimated with second degree polynomials,

$$\begin{bmatrix} x \\ y \\ z \end{bmatrix} = \begin{bmatrix} w_{2x}t^2 + w_{1x}t + w_{0x} \\ w_{2y}t^2 + w_{1y}t + w_{0y} \\ w_{2z}t^2 + w_{1z}t + w_{0z} \end{bmatrix}, \tag{8.14}$$

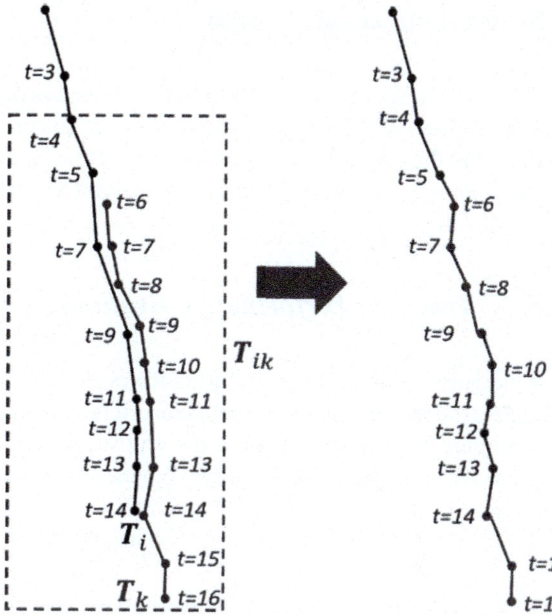

Fig. 8.15 Schematics of trajectory fusion

with variable t and coefficient vector $[w_{2x}, w_{1x}, w_{0x}, w_{2y}, w_{1y}, w_{0y}, w_{2z}, w_{1z}, w_{0z}]$. Considering that all particles are blown via a fixed lance with a simplified spatial coordinate $[x_0, y_0, z_0]$, we express the polynomial regression model as

$$\begin{bmatrix} x \\ y \\ z \end{bmatrix} = \begin{bmatrix} w_{2x}t^2 + w_{1x}t + x_0 - w_{2x}t_0^2 - w_{1x}t_0 \\ w_{2y}t^2 + w_{1y}t + y_0 - w_{2y}t_0^2 + w_{1y}t_0 \\ w_{2z}t^2 + w_{1z}t + z_0 - w_{2z}t_0^2 + w_{1z}t_0 \end{bmatrix}, \tag{8.15}$$

with

$$\begin{bmatrix} x_0 \\ y_0 \\ z_0 \end{bmatrix} = \begin{bmatrix} w_{2x}t_0^2 + w_{1x}t_0 + w_{0x} \\ w_{2y}t_0^2 + w_{1y}t_0 + w_{0y} \\ w_{2z}t_0^2 + w_{1z}t_0 + w_{0z} \end{bmatrix}. \tag{8.16}$$

The constraint, viz. that all particle trajectories pass through the point $[x_0, y_0, z_0]$, results in a polynomial model with the following coefficient parameters: $w_{2x}, w_{1x}, w_{2y}, w_{1y}, w_{2z}, w_{1z}, t_0$, whereby t_0 represents the time when a particle exactly leaves the lance. The polynomial fitting will be accomplished by solving least squares utilizing the trust region approach [21].

8.5 Experimental Results

In this section, we introduce an approach to assess tracking performance developed in [19]. Besides, the utilized parameters, the proposed 2.5D multiple particle tracking results, and the post-processing are also described and evaluated. Finally, polynomial regression of the obtained 3D particle trajectories is implemented.

8.5.1 Tracking Performance Evaluation Approach

For the purpose of tracking performance evaluation, [19] developed an approach to calculate the Jaccard similarity between reference tracks and candidate tracks. A track is considered as a series of subsequent positions $\{(x(t), y(t))\}$, with $t = t_0 \dots t_{end}$. The Euclidian distance between two positions $\theta_1(t)$ and $\theta_2(t)$ is defined as:

$$\|\theta_1(t) - \theta_2(t)\|_{2,\epsilon} \triangleq \min\big(\|\theta_1(t) - \theta_2(t)\|_{2}, \epsilon\big). \quad (8.17)$$

Here, ϵ denotes the gate. If two positions are more than ϵ apart, they will be considered as an impossible matching pair. Based on the distance between two positions, the distance between two existing tracks is formulated as:

$$d(\theta_1 - \theta_2) = \sum_{t=0}^{T-1} \|\theta_1(t) - \theta_2(t)\|_{2,\epsilon}, \quad (8.18)$$

where T represents the length of the image sequence.

Assuming a set of ground truth tracks denoted as $\mathcal{X}$ with $|\mathcal{X}|$ tracks and a set of candidate tracks $\mathcal{Y}$ with $|\mathcal{Y}|$ tracks, the distance between $\mathcal{X}$ and $\mathcal{Y}$ is defined as:

$$d(\mathcal{X}, \mathcal{Y}) = \min \sum_{k=1}^{|\mathcal{X}|} d(\theta_{\mathcal{X}}^k, \theta_{\mathcal{Z}}^k),\ \mathcal{Z} \in \Omega_{\mathcal{Y}}. \quad (8.19)$$

$\Omega_{\mathcal{Y}}$ denotes the obtained ensemble sets of tracks by choosing $|\mathcal{X}|$ tracks from the extended candidate tracks $\widetilde{\mathcal{Y}}$, which extends $\mathcal{Y}$ with $|\mathcal{X}|$ dummy empty tracks. $\mathcal{Z}$ represents an arbitrary element in $\Omega_{\mathcal{Y}}$. The assignment problem between $\mathcal{X}$ and $\widetilde{\mathcal{Y}}$ is addressed by using the Munkres algorithm.

By finding the optimal pairing $(\mathcal{X}, \mathcal{Z}^*)$ with minimum distance, the following indices to assess the tracking performance are computed:

(1) Normalized pairing score

$$\alpha(\mathcal{X}, \mathcal{Y}) \triangleq 1 - d(\mathcal{X}, \mathcal{Y})/d(\mathcal{X}, \varnothing). \quad (8.20)$$

$\varnothing$ stands for the set of $|\mathcal{X}|$ dummy empty tracks.

(2) Full normalized score

$$\beta(\mathcal{X}, \mathcal{Y}) \triangleq \frac{d(\mathcal{X}, \varnothing) - d(\mathcal{X}, \mathcal{Y})}{d(\mathcal{X}, \varnothing) + d(\overline{\mathcal{Y}}, \varnothing_{\overline{\mathcal{Y}}})}, \tag{8.21}$$

where $\overline{\mathcal{Y}}$ denotes the tracks in $\mathcal{Y}$ but not in $\mathcal{Z}^*$ and $\varnothing_{\overline{\mathcal{Y}}}$ stands for a set of dummy tracks with $|\overline{\mathcal{Y}}|$elements.

(3) The Jaccard similarity index for positions

$$JSC \triangleq \frac{TP}{TP + FN + FP}. \tag{8.22}$$

If the distance between two temporal positions is smaller than ϵ, they are considered as a matching pair. The number of matching pairs of positions in $(\mathcal{X}, \mathcal{Z}^*)$ is denoted as TP. FN stands for the number of non-matching pairs. FP represents the total amount of positions in $\overline{\mathcal{Y}}$.

(4) The Jaccard similarity index for tracks

$$JSC_\theta \triangleq \frac{TP_\theta}{TP_\theta + FN_\theta + FP_\theta}. \tag{8.23}$$

TP_θ indicates the number of non-dummy tracks in $\mathcal{Z}^*$. On the contrary, FN_θ denotes the number of dummy tracks in $\mathcal{Z}^*$. FP_θ counts the number of tracks in $\overline{\mathcal{Y}}$.

8.5.2 Parameterization and Results of 2.5D Particle Tracking

The entailed threshold values recommended in Sect. 8.3 are given in Table 8.1. Despite a large number of values, they are easy to determine due to a statistical analysis of the obtained particle tracklets.

Table 8.1 Parameter value

Parameter	Value	Unit	Parameter	Value	Unit
D_ϵ	2	frame	$t_{\epsilon,1}$	15	frame
$\lvert\overline{V}\rvert_\epsilon$	1	pixel/ frame	$t_{\epsilon,2}$	5	frame
R_ϵ	0.9	–	z_ϵ	1500	mm
$\lvert\Delta V\rvert_\epsilon$	18	pixel/ frame	ϵ	10	pixel
θ_ϵ	30	°			

To comprehensively investigate the flight behavior of a specific fuel particle type, we utilize only one variety of particles for each experiment. There are four varieties of particles overall, viz. wood chips, confetti from a paper punch, paper shreds from a shredder, and PE granules, as illustrated in Fig. 8.16. The particle sizes can be estimated according to the 1 cm scale in the bottom right of each image. Since the fuel particles are relatively small, they could be "invisible" sometimes, enhancing particle tracking difficulty. Moreover, particles of the same variety appear similarly, which leads to a low possibility of extracting object properties for detection and tracking.

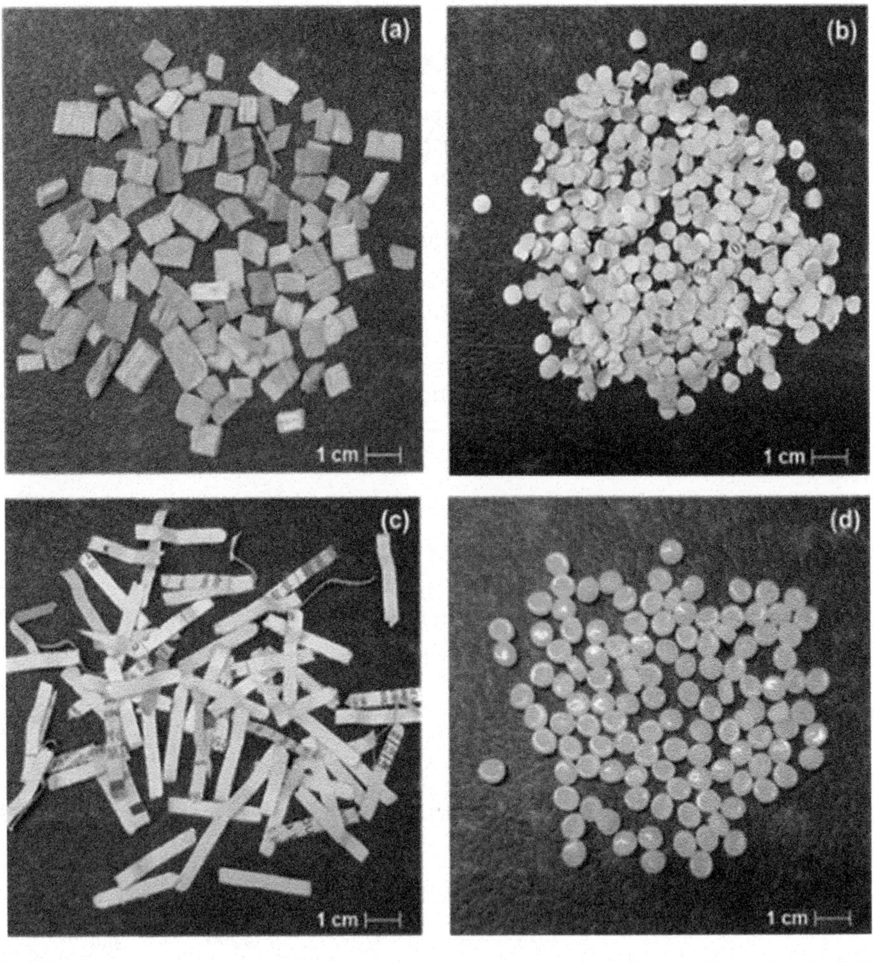

Fig. 8.16 Various refuse-derived fuel particles. **a** Wood chips. **b** Confetti. **c** Paper shreds. **d** PE granules

Because of the lack of adequate accuracy of the light-field camera calibration at the image edges, we define a region of interest (ROI) before particle detection, as depicted in Fig. 8.17. Particles outside of the ROI are ignored.

The quantitative evaluations of tracking results acquired by the proposed approach without and with post-processing are registered in Tables 8.2 and 8.3, respectively. As proved by the tables, the proposed post-processing significantly facilitates the

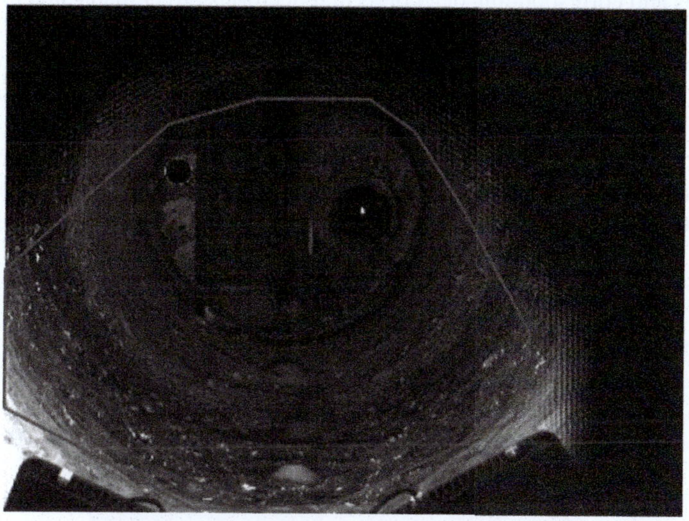

Fig. 8.17 Region of Interest (blue marked)

Table 8.2 Evaluation of tracking performance

Fraction	Measure				
	d	α	β	JSC	JSC_θ
Wood chips	4412.49	0.51	0.12	0.17	0.02
Confetti	17,583.83	0.29	0.14	0.18	0.08
Paper shreds	15,907.52	0.22	0.1	0.13	0.02
PE granules	5049.27	0.45	0.28	0.34	0.06

Table 8.3 Evaluation of tracking performance with post-processing

Fraction	Measure				
	d	α	β	JSC	JSC_θ
Wood chips	2778.32	0.69	0.69	0.79	1.0
Confetti	13,928.38	0.44	0.40	0.44	0.81
Paper shreds	10,743.83	0.48	0.45	0.56	0.67
PE granules	3830.96	0.59	0.59	0.66	0.92

tracking performance with respect to the entire five measures. Since a lot of faulty tracklets and outliers in the tracklets are deleted, the Jaccard similarity index for positions (JSC) and tracks (JSC_θ) achieve considerable enhancements. Meanwhile, the matching distance (d) and pairing scores (α, β) are meliorated as well. Nevertheless, some parts of tracklets are discarded owing to the ROI, and hence, JSC remains lower than JSC_θ overall. Moreover, several particles are at times much larger than ϵ due to the exposure time, which results in the distances between ground truth detections and corresponding detections in trajectories larger than ϵ and thus, wrong pairing of detections and consequently lower JSC.

The tracking performances of wood chips and PE granules are more precise than the other two in terms of the five measures. Since the wood chips and PE granules are so heavy and small that they mainly fall down speedily without apparent rotations, their trajectories are distinct and clear for tracking. Figure 8.18 shows an example of the tracking result of PE granules within 150 frames. Even before the post-processing, several complete trajectories are already acquired. As a consequence of implementing

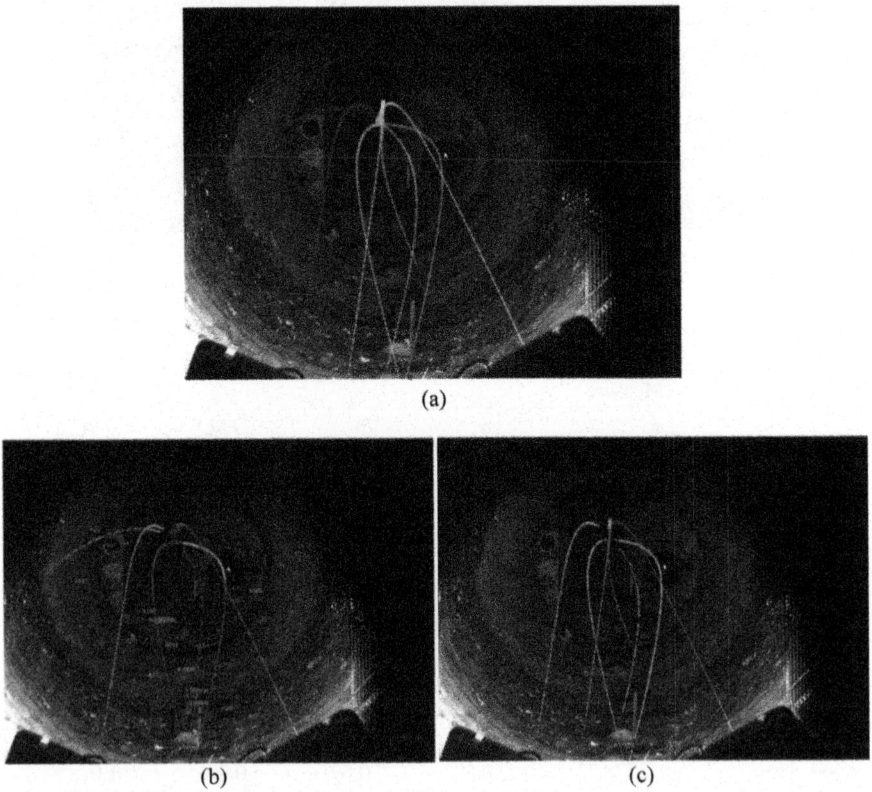

Fig. 8.18 Tracking result of PE granules. Different trajectories are marked with different clolors. **a** Manually identified ground truth. **b** Tracking result without post-processing. **c** Tracking result with post-processing

Fig. 8.19 Missed trajectory (red marked, trajectory in dark blue). Left: marked trajectory area in an added image. Right: visual presentation of the corresponding ground truth

the post-processing, the tracking result with post-processing shows a strong similarity to the ground truth despite some missing detections. However, one trajectory is entirely ignored due to the small size and low illumination of the corresponding particle. The trajectory is marked in Fig. 8.19, which is too small and dark to identify optically.

The tracking performances of confetti and paper shreds are relatively worse than wood chips and PE granules. Small paper is in comparison to wood and PE light in weight and therefore falls slowly down with rotations in the air, especially for paper shreds with a sizeable windward surface. Hence, their trajectories are tangled in the majority of cases, which exerts adverse effects on particle tracking. Figure 8.20, for instance, illustrates an example of the tracking result of confetti within 230 frames. Compared to the result of PE granules, the trajectories are more complicated and irregular. Because the confetti moves slowly, tracklets 11 and 15 are failed to be connected due to a time gap longer than the threshold value.

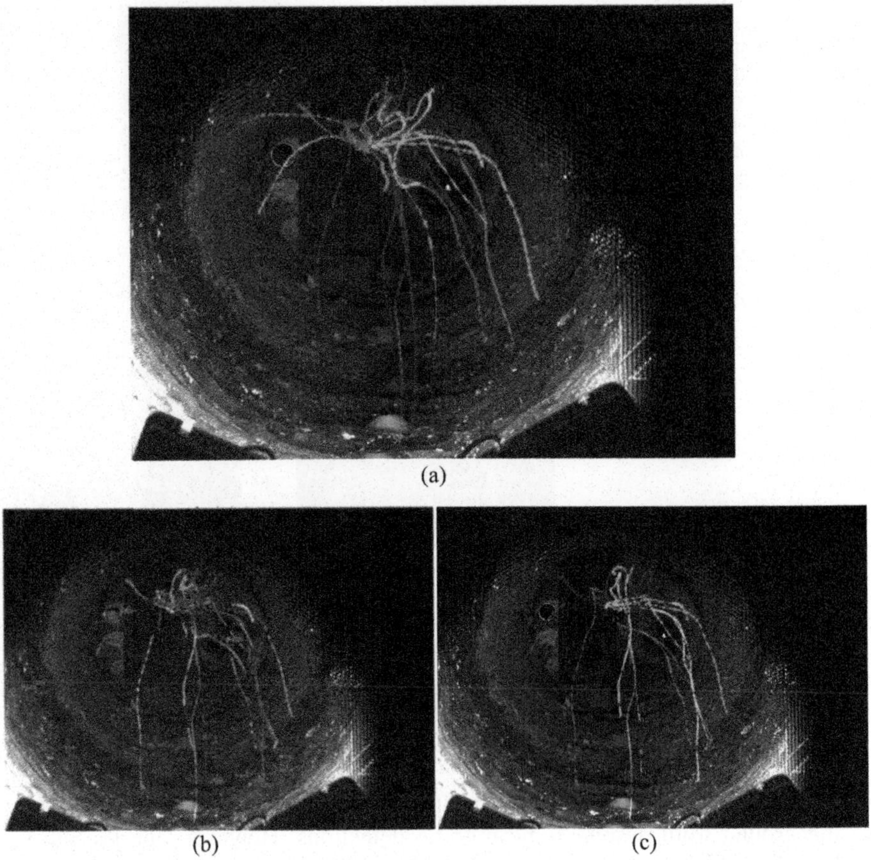

Fig. 8.20 Tracking result of confetti. a Manually identified ground truth. b Tracking result without post-processing. c Tracking result with post-processing

8.5.3 Results of 3D Particle Trajectories Regression

As mentioned before, we are able to extract 3D particle trajectories based on the obtained 2D trajectories. The directly acquired jectories are generally fluctuated caused by the inaccuracy of the measurement and therefore require regressions for further analysis. We estimate the trajectories with second degree polynomials at first and increase the degree to third if the estimation performance is insufficient. The obtained regression model are assessed in accordance with the measures elaborated in [20], viz. root mean squared error (RMSE), mean absolute percentage error (MAPE), R-squared R^2, and adjusted R-squared R^*. Table 8.4 documents the evaluation of the trajectories of various fuel particles.

Table 8.4 Results of 3D particle Trajectories regression

Fraction	Measure				
	Degree x y z	RMSE x y z	MAPE x y z	R^2 x y z	R^* x y z
Wood chips	2 3 3	11.39 10.90 222.41	14.49 13.64 1.69	0.97 0.93 0.76	0.97 0.92 0.68
Confetti	3 3 3	13.79 14.82 269.91	9.61 17.24 2.48	0.90 0.92 0.47	0.88 0.91 0.41
Paper shreds	3 3 3	10.94 10.80 185.88	3.81 9.35 1.93	0.90 0.91 0.58	0.85 0.87 0.47
PE granules	3 2 3	6.67 15.67 171.74	2.56 9.02 1.47	0.97 0.97 0.94	0.97 0.97 0.94

Since the fluctuation in the depth direction is the intensest, RMSE in the z-direction is significantly larger than the other two directions. In contrast, MAPE in the z-direction is the lowest caused by relatively large depth coordinates of the particle. In general, R^2 and R* in the x- and y- direction reach roughly 0.9, indicating the satisfied adequacy of the regression model. R^2 and R^* in the z-direction are with small values due to the fluctuations. Figure 8.21 depicts the estimated 3D trajectories of the wood chips, according to which the fluctuations in the depth direction are considerable and noticeable.

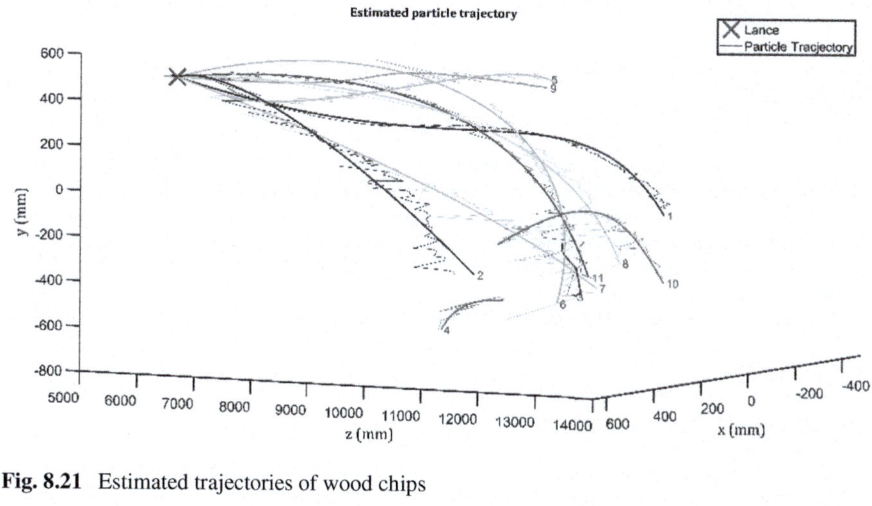

Fig. 8.21 Estimated trajectories of wood chips

8.6 Conclusion

This paper presented approaches to accurately analyze flight behaviors of various fuel particles based on images captured by a light-field camera. First of all, the proposed detection method identifies the particles, which combines the detection results of 2D SIFT and 3D Clustering method DBSCAN. The combined detection method is proved to achieve superior performance. Consequently, we utilized the linear Kalman filter and 2.5D GNN to estimate temporary particle positions and associate them with de-tected particles simultaneously afterward. Since the initial tracking results contained plenty of faulty and incomplete tracklets, we developed a framework for tracking post-processing that takes full advantage of the motion similarity between tracklets. With the proposed post-processing method for Multiple-Particle-Tracking, we were able to not only correct the trajectories but also facilitate the selection of the tracking parameters. Since faulty or incomplete tracklets caused by impropriate parameters are pruned or modified, precise final trajectories can be obtained, even if tracking parameters, as cost of none assignment or maximum frames for none assignment, are not chosen optimally. Based on the 2D trajectories, we were capable of extracting 3D particle trajectories and estimating them with polynomials afterward.

Our experiments with various fuel particles indicate the accuracy and adequacy of the proposed approach for post-processing. Future researches will aim at the robustness of the post-processing approach for wide applications.

Acknowledgements This study is supported by AiF – German Federation of Industrial Research Associations (No. 20410N).

Notes The paper was originally presented in the *4th International Joint Conference on Computer Vision and Pattern recognition (CCVPR2021)* and extended to be published in the proceeding of the *2021 International Symposium on Automation, Mechanical and Design Engineering (SAMDE 2021)*.

References

1. Otsu, N.: A threshold selection method from grey level histograms. In: IEEE Transactions on Systems, Man, and Cybernetics. New York, S. pp. 62–66. ISSN 1083–4419, 9 (1979)
2. Lowe, D.G.: Object recognition from local scale-invariant features. In: Proceedings of the Seventh IEEE International Conference on Computer Vision, vol. 2, pp. 1150–1157 (1999)
3. Lowe, D.G.: Distinctive image features from scale-invariant keypoints. Int. J. Comput. Vision **60**, 91–110 (2004)
4. Ouyang, W. et al.: DeepID-Net: Deformable deep convolutional neural networks for object detection. In: 2015 IEEE Conference on Computer Vision and Pattern Recognition (CVPR), pp. 2403–2412 (2015)
5. Szegedy, C., Toshev, A., Erhan, D.: Deep Neural Networks for Object Detection (2013)
6. Cheng, Y.: Mean shift, mode seeking, and clustering. IEEE Trans. Pattern Anal. Mach. Intell. **17**(8), 790–799 (1995)
7. Ester, M., Kriegel, H.-P., Sander, J., Xu, X.: A density-based algorithm for discovering clusters in large spatial databases with noise. In: Proceedings of 2nd International Conference on Knowledge Discovery and Data Mining 96

8. Zhang, M., Matthes, J., Aleksandrov, K., Gehrmann, H.-J., Vogelbacher, M.: Detection and tracking of refuse-derived fuel particles based on a light field camera system. Technisches Messen **88**(6), 396:407 (2021)
9. Kalman, R.E.: A new approach to linear filtering and prediction problems. J. Basic Eng. **82**, 35–45 (1960)
10. Einicke, G.A., White, L.B.: Robust extended Kalman filtering. IEEE Trans. Signal Process. **47**(9), 2596–2599 (1999). September
11. Julier, S.J., Uhlmann, J.K.: Unscented filtering and nonlinear estimation. Proceedings of the IEEE **92**(3), 401–422 (2004). March
12. Konstantinova, P., Udvarev, A., Semerdjiev, T.: A study of a target tracking algorithm using global nearest neighbor approach. In: international Conference on Computer Systems and Technologies (2003)
13. Bar-Shalom, Y., Tse, E.: Tracking in a cluttered environment with probabilistic data association. Automatica **11**(5), 451–460. ISSN 0005–1098 (1975)
14. Fortmann, T.E., Bar-Shalom, Y., Scheffe, M.: Multi-target tracking using joint probabilistic data association. In: 19th IEEE Conference on Decision and Control including the Symposium on Adaptive Processes, pp. 807–812 (1980)
15. Mušicki, D., Evans, R.: Joint integrated probabilistic data association: JIPDA. IEEE Trans. Aerosp. Electron. Syst. **40**(3), 1093–1099 (2004). July
16. Reid, D.: An algorithm for tracking multiple targets. IEEE Trans. Autom. Contr. **24**(6), 843–854 (1979). December
17. Ma, C., Yang, C., Yang, F., Zhuang, Y., Zhang, Z., Jia, H., Xie, X.: Trajectory factory: tracklet cleaving and re-connection by deep siamese Bi-GRU for multiple object tracking. In: 2018 IEEE International Conference on Multimedia and Expo (ICME), pp. 1–6 (2018)
18. Peng, J., Wang, T., Lin, W., Wang, J., See, J., Wen, S., Ding, E.: TPM: multiple object tracking with tracklet-plane matching. Pattern Recognit. **107**, 107480. ISSN 0031-3203 (2020)
19. Chenouard, N., Smal, I., de Chaumont, F., et al.: Objective comparison of particle tracking methods. Nat Methods **11**, 281–289 (2014)
20. Ostertagová, E.: Modelling using polynomial regression. Procedia Eng. **48**, 500–506. ISSN 1877-7058 (2012)
21. Coleman, T.F., Li, Y.: An interior, trust region approach for nonlinear minimization subject to bounds. SIAM J. Optim. **6**, 418–445 (1996)

Chapter 9
Analysis of the Combustion Process of a Diesel Engine Fuelled by Diesel and Hydrogen

Rafał Longwic, Gracjana Woźniak Borawska, Dawid Tatarynow, Przemysław Sander, and Arkadiusz Gita

Abstract This article discusses the results of studies on the use of hydrogen additive for combustion together with diesel fuel in a compression ignition engine. A motor with Common Rail injection system was tested. Hydrogen was administered sequentially into the engine intake channel. The engine fitted to the vehicle was inducted and the engine load was triggered using a two-axle chassis dynamometer. A proprietary hydrogen injection system was used during the preliminary testing phase. Energy and environmental performance of the engine built into the vehicle was recorded. An analysis of the effect of the hydrogen additive on the above-called parameters was made. The proposed solution may contribute to the longer service life of current compression ignition engines due to improved ecological properties of such sources of propulsion.

Keywords Hydrogen · Diesel fuel · Combustion engine · Environmental parameters · Emission · Engine induction system AVL

9.1 Introduction

Internal combustion engines are still one of the basic solutions among the sources of propulsion of machines and motor vehicles used, which currently makes them impossible to replace with other solutions. Among internal combustion engines, the most popular engine is the compression ignition engine especially in heavy goods vehicles, sea-vessels, and passenger vehicles. It is estimated that more than 30% of passenger cars and almost 93% of vans were sold with diesel in the European Union in 2019 [1]. The high efficiency of the compression ignition engine and the availability of petroleum fuels means that this source of propulsion will be used for many years despite the trend away from fossil fuels. In order to extend the life of the compression ignition engine, the possibility of reducing fuel consumption and alternative low-emission fuels is sought in order to reduce emissions harmful

R. Longwic (✉) · G. W. Borawska · D. Tatarynow · P. Sander · A. Gita
Faculty of Mechanical Engineering, Lublin University of Technology, 22-618 Lublin, Poland
e-mail: r.longwic@pollub.pl

M. A. Laribi et al. (eds.), *Advances in Automation, Mechanical and Design Engineering*, Mechanisms and Machine Science 121,
https://doi.org/10.1007/978-3-031-09909-0_9

to the environment [2, 3]. One way to improve the performance and environmental performance of engines is to use hydrogen as a fuel. Hydrogen as an alternative to traditional conventional fuels can be used to power vehicles as fuel in traditional engines and fuel cells [4]. The Hydrogen Council predicts that by 2050, around 20–25% of passenger traffic will be served by hydrogen-powered vehicles. Data presented in the document "Hydrogen—scaling up. A sustainable pathway for the global energy transition" shows that hydrogen could become a fuel for more than 400 million cars, 15–20 million trucks and about 5 million buses by 2050 [5].

The fact that its stocks are practically inexhaustible is a feature of hydrogen. It can be sourced from both non-renewable and renewable sources. Among the methods of hydrogen extraction are steam reforming, pyrolysis, coal gasification, biomass gasification, gasoline reforming, and electrolysis. However, the use of hydrogen also entails technological challenges. One of them, resulting from low hydrogen density, is its adequate storage especially in vehicles where it poses a significant risk of use. In addition, low ignition energy (0.02 mJ) can cause premature ignition, which can lead, among other things, to the formation of nitrogen oxides, a decrease in efficiency or uneven engine operation. In the case of hydrogen, and in particular its wide flammability limits in the range of 4 to 75% of the make it perfectly combined with other fuels [4, 6, 7].

Research [4, 6, 8–13] on the use of hydrogen as a fuel or additive to existing fuels is currently being conducted worldwide. These are studies showing the possibilities resulting from its use in the combustion process, in particular by reducing exhaust emissions or increasing combustion efficiency and engine life. It is assumed that the hydrogen additive may reduce emissions of carbon monoxide (CO), hydrocarbons (HC) and carbon dioxide (CO2) [14]. In addition, an increase in NOx can be observed in the case of hydrogen [6]. The aim is therefore to seek solutions and methods for reducing NOx.

9.2 Test Bench and Methodology

The research was carried out at the Centre for Innovation and Technology Transfer of the Lublin University of Technology, in the Laboratory of the Department of Motor Vehicles, equipped with, among others, the Dynorace chassis dynamometer, designed for vehicles with two axles type DF4FS-HLS. The test bench is shown in Fig. 9.1. The detailed construction of the position and the parameters of the vehicle to be tested are described in the authors' publication [15]. Prototype gaseous fuel dispensing systems were used for the study. The concentration of toxic compounds in the exhaust gas using the MAHA exhaust gas analyser was measured. The engine was inducted using AVL's INDIMICRO 602 engine induction system. AVL signals are the pressure flow inside the cylinder, the signal of the engine crankshaft position, the injection pressure [15]. A diagnostic interface designed to work with FIAT ECU SCAN was used to analyse the operating parameters of the engine controller.

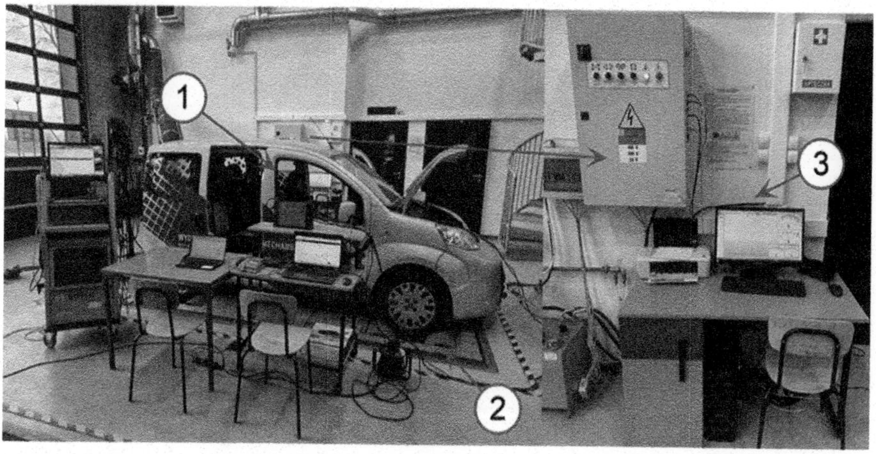

Fig. 9.1 Test station for diesel and hydrogen powered vehicle: (1) Fiat Qubo tested vehicle with 1.3 Multijet engine, equipped with AVL's INDIMICRO 602 engine abdication system and MAHA exhaust gas analyser. (2) Chassis dynamometer DF4FS-HLS. (3) Computer with DynoRace software installed, including control cabinet and sensor accessories

The tests were carried out under the driving conditions of the vehicle on the chassis dynamometer, with the vehicle being subjected to rolling resistance. They were intended to determine the effect of the hydrogen additive on the performance of the compression ignition engine. The measurements were carried out on the 4th gear at the fixed engine speeds of 1500, 2000, 2500, 3000, 3500, 4000, 4500 rpm, for two engine supply variants, i.e. diesel fuel, and for diesel fuel with gas-injected hydrogen into the engine intake manifold. Hydrogen was administered sequentially according to ignition order 1-3-4-2 when the suction valves opened on subsequent cylinders. The hydrogen pressure ranged from 2 to 2.5 bar, while the opening time of hydrogen injectors was 2 and 3 ms. The opening time of the injectors was constant for all rotational speeds. Atmospheric pressure during the studies was 978 hPa.

9.3 Results and Conclusions

9.3.1 Results

An analysis of the results of the studies shows that:

- with additional hydrogen supply, it was noted that the amount of diesel injected was reduced, for the opening time of the hydrogen injector 2 ms the amount of injected diesel fuel was from 0.3 to 5.0 mm^3 per cycle, and for 3 ms from 0.1 to 4.8 mm^3 per cycle, the biggest difference was noticed at engine speed of 1500 rpm;

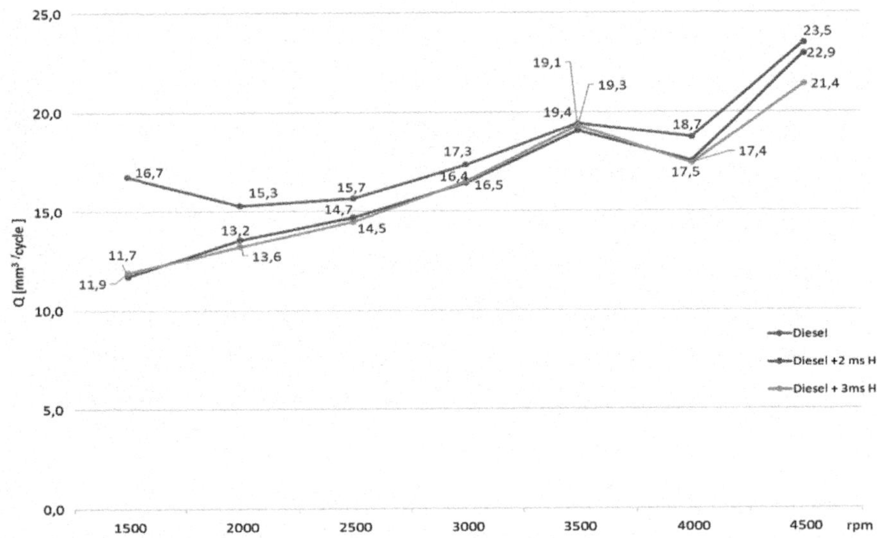

Fig. 9.2 Amount of injected fuel depending on the rotational speed at the supply of diesel and diesel with the addition of hydrogen for different injector opening times

- it has been observed that for an engine speed of 3500 rpm, the addition of hydrogen regardless of the opening time of the injectors slightly reduces the amount of injected diesel fuel, the corresponding results are shown in Fig. 9.2.

In view of the above, the hydrogen additive affects the angular tilt positions of the vehicle acceleration lever. The engine speed of 1500 rpm was achieved at a tilt value of 8.6° (this is 59.1% less than on diesel where the tilt was 27.8°), for the engine speed 2000 rpm the difference was 13%, for subsequent speeds the above difference had a decreasing trend and was 4%–1%. The recorded average pressure predicted over the entire engine speed range remained at a lower level than that of diesel combustion alone (Fig. 9.3), the largest difference at 1500 rpm was 22–25%. The maximum combustion pressure was also slightly lower while burning diesel and hydrogen. The maximum difference occurred at 1500 rpm difference and was 12–14% (Fig. 9.4). In the case of the maximum rate of build-up of combustion pressure, it was noted that at the hydrogen injection times used, the pressure increased more slowly at speeds between 1500 and 2500 rpm, while at further measuring points the values were similar to those for diesel.

Toxicity measurements were made before the catalyst. NOx for the first two rotational speeds tended to decrease by around 59% compared to diesel. At subsequent rotational speeds they had comparable values (Fig. 9.5). In the case of particulate matter, a significant decrease was observed compared to the combustion of diesel fuel alone (Fig. 9.6). In the case of NO at speeds of 1500 and 2000 rpm, a decreasing trend was also noted (Fig. 9.7). A slight decrease in carbon dioxide concentration (Fig. 9.8) was observed for all rotational speeds and all injector opening times.

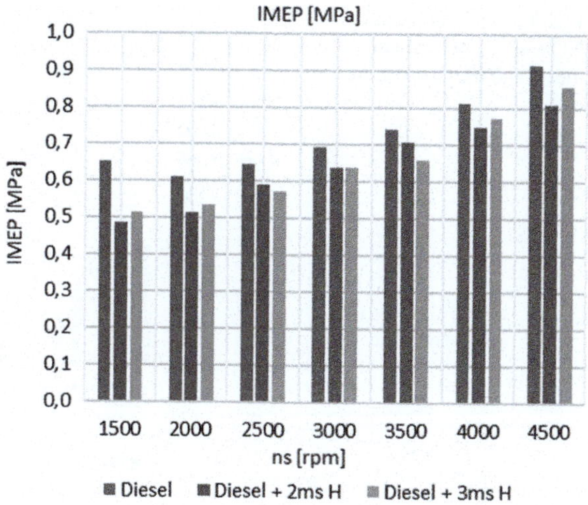

Fig. 9.3 Indicated mean pressure effective pressure (IMEP) depending on the rotational speed at the supply of diesel and diesel with the addition of hydrogen for different injector opening times

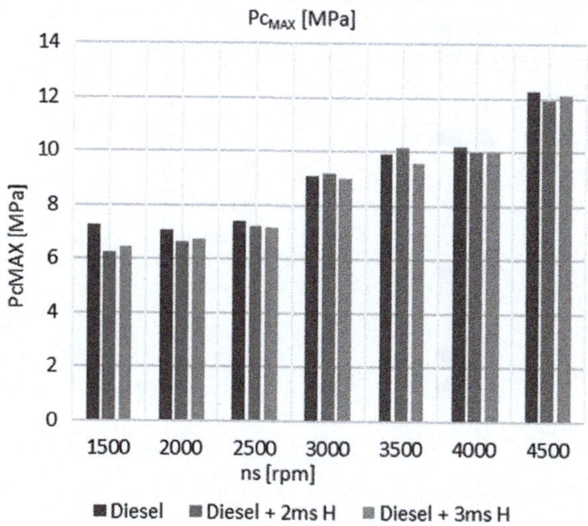

Fig. 9.4 Maximum combustion pressure depending on the rotational speed at the supply of diesel and diesel with the addition of hydrogen for different injector opening times

9.3.2 Conclusions

The article shows preliminary results of tests on the simultaneous use of hydrogen and diesel to power the diesel engine of a traction vehicle. A prototype hydrogen sequential injection plant for the intake channel was used. Promising results were achieved in terms of energy and environmental performance of the engine. The toxicity was measured before the catalyst. It was noted that the injection of hydrogen reduces the amount of air drawn into the combustion system, so it is necessary to affect the diesel dosage at the same time (Fig. 9.2). In Fig. 9.5 it is observed that

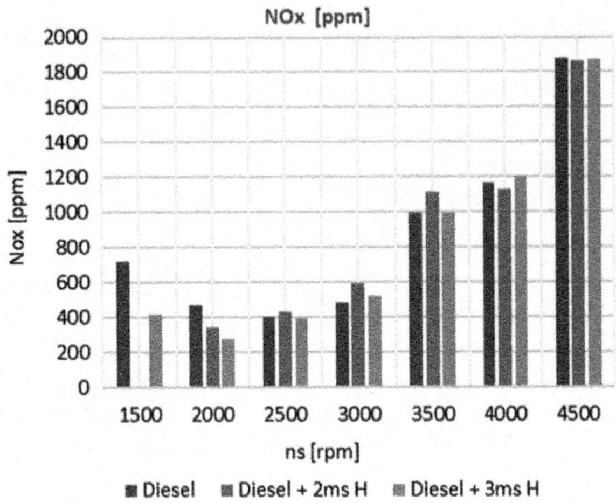

Fig. 9.5 Concentration of NOx when fuelled with diesel and diesel-hydrogen

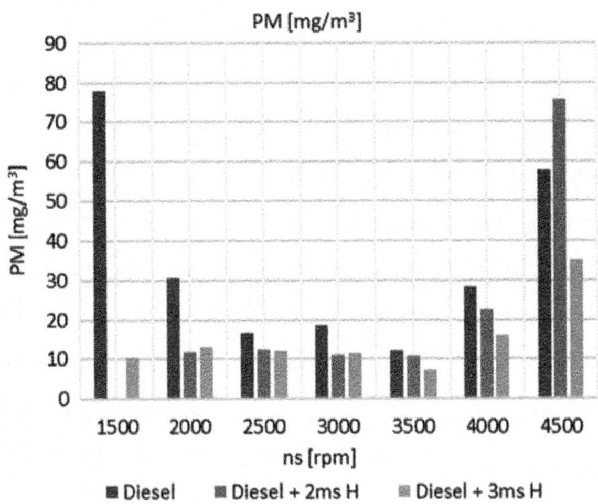

Fig. 9.6 Concentration of PM when fuelled with diesel and diesel-hydrogen

hydrogen addition reduced NOx at 1500 and 2000 rpm. Figure 9.6 illustrates that the addition of hydrogen causes a reduction in particulate matter emissions. A similar relationship to NOx was noted in the recorded NO values (Fig. 9.7). A slight decrease in carbon dioxide concentration (Fig. 9.8) was observed for all rotational speeds and all injector opening times. The presented results show the possibility of implication of the prototype installation to extend the life of compression ignition engines. Further studies will focus on optimising the amount of hydrogen administered due to the combustion process.

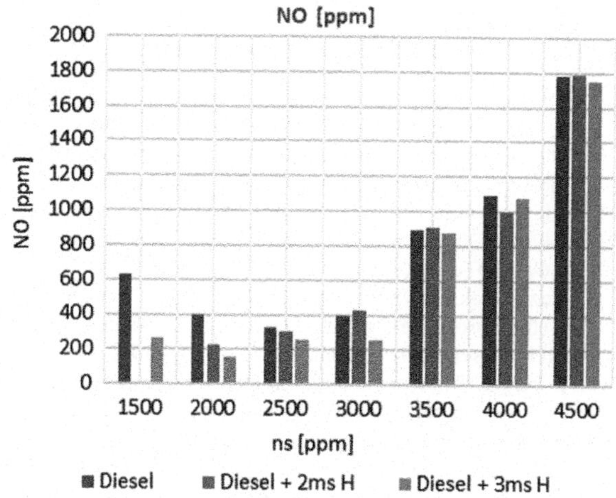

Fig. 9.7 Concentration of nitrogen oxide on when fuelled with diesel and diesel-hydrogen

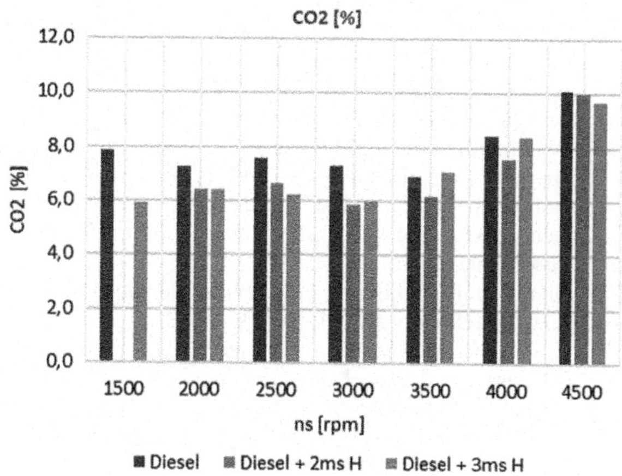

Fig. 9.8 Concentration of carbon dioxide when fuelled with diesel and diesel-hydrogen

Funding Sources This project/research was financed in the framework of the project Lublin University of Technology-Regional Excellence Initiative, funded by the Polish Ministry of Science and Higher Education (contract no. 030/RID/2018/19).

References

1. European Automobile Manufacture Association, Polski Związek Przemysłu Motoryzacyjnego: W kierunku zeroemisyjnej mobilności. Czynniki determinujące rozwój napędów alternatywnych w samochodach osobowych i dostawczych w Unii Europejskiej. Raport (2020)
2. Górski, K., Sander, P., Longwic, R.: The assessment of ecological parameters of diesel engine supplied with mixtures of canola oil with n-hexane. IOP Conf. Ser.: Mater. Sci. Eng. **421**(417), 042025 (2018). In: Scientific Conference on Automotive Vehicles and Combustion Engines, KONMOT 2018, Cracow, 13–14 September 2018, p. 141575
3. Zdziennicka, A., Szymczyk, K., Janczuk, B., Longwic, R., Sander, P.: Surface, volumetric, and wetting properties of oleic, linoleic, and linolenic acids with regards to application of canola oil in diesel engines. Appl. Sci. (Switz.) **9**(171), 3445 (2019)
4. Longwic, R., Woźniak, G., Sander, P.: Compression-ignition engine fuelled with diesel and hydrogen engine acceleration process. Combust. Engines **180**(1), 47–51 (2020)
5. Feri: Energetyka zrównoważona. WODÓR (2018)
6. Hamdan, M.O., Selim, M.Y.E., Al-Omari, S.-A.B., Elnajjar, E.: Hydrogen supplement co-combustion with diesel in compression ignition engine. Renew. Energy **82**, 54–60 (2015)
7. Verhelst, S., Wallner, T.: Hydrogen-fueled internal combustion engines. Prog. Energy Combust. Sci. **35**(6), 490–527 (2009)
8. Heffel, J.W.: NOx emission reduction in a hydrogen fueled internal combustion engine at 3000 rpm using exhaust gas recirculation. Int. J. Hydrog. Energy **28**(11), 1285–1292 (2003)
9. Masood, M., Ishrat, M.M., Reddy, A.S.: Computational combustion and emission analysis of hydrogen–diesel blends with experimental verification. Int. J. Hydrog. Energy **32**(13), (2007)
10. Dimitrov, E., Gigov, B., Pantchev, S., Michaylov, P., Peychev, M.: A study of hydrogen fuel impact on compression ignition engine performance MATEC. Web Conf. **234**, (2018)
11. Parlak, A., Ayhan, V., Üst, Y., Sahin, B., Cesur, İ., Boru, B., Kokkulunk, G.: New method to reduce NOx emissions of diesel engines: electronically controlled steam injection system. J. Energy Inst. (2012)
12. Dimitriou, P., Tsujimura, T.: A review of hydrogen as a compression ignition engine fuel. Int. J. Hydrog. Energy **42**(38), (2017)
13. Singhyadav, V., Soni, S.L., Sharma, D.: Performance and emission studies of direct injection C.I. engine in dual fuel mode (hydrogen-diesel) with EGR. Hydrog. Energy (2012)
14. Karagoz, Y., Sandalci, T., Yuksek, L., Dalkilicx, A.S., Wongwises, S.: Effect of hydrogen–diesel dual-fuel usage on performance, emissions and diesel combustion in diesel engines. Adv. Mech. Eng. (2016)
15. Longwic, R., Sander, P., Zdziennicka, A., Szymczyk, K., Jańczuk, B.: Combustion process of canola oil and n-hexane mixtures in dynamic diesel engine operating conditions. MDPI (2020)

Chapter 10
Comparative Study of MPC and PID Controllers in Autonomous Vehicle Application

Sara Abdallaoui, Ali Kribèche, and El-Hassane Aglzim

Abstract The main challenging task in autonomous navigation is path-following control where the vehicle required to track a given path. Which can be interpreted by the error between the reference path and the desired one is to converge to zero and it gets more complicated when the vehicle dynamics is considered, such as non-linear vehicle dynamics, fast sampling time and limited computing resources. In this paper, we compare and describe the performance of the control law based on PID and MPC controllers for vehicle path tracking. PID controller is used for controlling the speed, stability and precision of the response, and eliminating the static error. MPC controller is used because of its known advantages such as the explicit handling of constraints explicit use of a model, well understood and defined tuning parameters and the most interesting advantage is the ability to predict future behavior of the system and take it into consideration.

Keywords Model predictive control · Autonomous vehicle control · Model predictive control (MPC) · Proportional-integral-derivative (PID) · Path tracking

10.1 Introduction

Technology helps humans in their daily life, facilitates tasks, improves productivity and quality of life. One of the most popular topics is about autonomous vehicles. It gets more and more interest by researchers and tech companies in recent decade by the evolution of Machine learning, Deep learning, Control theory, Computer Vision, Signal Processing etc. Therefore, the application of this technology was applied first to manufacturing industries then it becomes widely used in fields of space, medicine, military, agriculture and so forth. Before the applications in real world machines, developers need to evaluate their studies and the behavior of the machine with the environment without exposing people to possible dangerous situations.

S. Abdallaoui (✉) · A. Kribèche · E.-H. Aglzim
DRIVE EA1859, Univ. Bourgogne Franche Comté, 58000 Nevers, France
e-mail: Sara.Abdallaoui@u-bourgogne.fr

M. A. Laribi et al. (eds.), *Advances in Automation, Mechanical and Design Engineering*, Mechanisms and Machine Science 121,
https://doi.org/10.1007/978-3-031-09909-0_10

Autonomous vehicles are expected to make our life easier and rather to provide cheaper and safer transportation. An autonomous vehicle by definition is a vehicle that is capable of sensing its environment and moving safely with little or no human input. A typical autonomous system is composed of the following functional modules: perception, prediction, planning and control. Perception is the capacity of detecting and tracking the objects of interest over time (e.g. other vehicles, pedestrians). The prediction module estimates the intentions and trajectories that the vehicle is to use to make decisions. Planning module finds out which trajectories are safe while control outputs the commands necessary to move the vehicle in such trajectory [1].

For autonomous vehicles, navigation is the most important factor in mobile robotics, it includes the process of identifying the robot's position simultaneously, planning the path, and following the path planned using data from on the most famous sensors for localization as GPS, Ultra-Wideband (UWB), and for navigation and detection, sensors as Lidar, Radar, RGB camera and event camera. The 3D object detection is the core of perception module. This system exploits Lidar or cameras data as input and provide 3D pose and dimension of obstacles. Thanks to several public datasets, the access to the development/test of many solutions become easy. It gives the knowledge of the best performance of all the methods at that moment. Today, new methods reach step by step to the very high accuracy but it is still not enough to meet some safety's constraint.

Researchers have been conducted on dynamic path planning and more based on how to increase the performance of these vehicles in such a difficult environment to avoid any strange situation that the vehicle may be able to confront in real driving. The main purpose of planning and navigation is to provide vehicles with a safe driving and collision-free path towards their destinations, accounting for vehicle dynamics, maneuvering capabilities in the presence of obstacles, and traffic rules and road boundaries. A global reasoning of all possible paths should finally allow finding the best safe path according to different criteria, such as distance, time, cost, difficulties … An intelligent planning system should be able to contemplate the nature, size, and the speed of the vehicle or the robot [2]. On the other side of autonomous driving, Neural Network algorithms (NNA) are widely used in autonomous mobile robot, not only in motion planning but also in robot perception, guidance, self-driving, obstacle avoidance and so forth. In dynamic path planning, Model Predictive Control (MPC) and Proportional-Integral-Derivative (PID) are widely used to ensure vehicle's tasks. MPC is a method used in the trajectory tracking of autonomous vehicles. It deals with both of linear and nonlinear systems and which requires online optimization problem. PID is a control loop mechanism that employ feedback control which is widely used in industrial control systems.

The remainder of the paper is organized as follows. Section 10.2 presents the vehicle lateral control mentioning several recent methods and applications used in the literature including our used approaches. A path tracking task for vehicles is introduced in Sect. 10.3. The two control approaches are presented in Sect. 10.4. Simulation and experimental results are given in Sect. 10.5 and comparing results in Sect. 10.6. Finally, a conclusion and presenting future work are drawn in Sect. 10.7.

10.2 Related Work

10.2.1 Vehicle Lateral Control

Autonomous driving can be divided into three stages: perception of the environment, trajectory planning and vehicle control. In this section we will provide the vehicle control methods and utility in path tracking. By definition vehicle control involves maneuvering the vehicle using actuators such as the steering wheel, brake and accelerator to ensure the reference path following. There are two ways to design controllers for autonomous driving; either based on the imitation of the human driver which includes approaches based on the use of driver models such as AI (Artificial Intelligence), or the use of approaches based on the theory command.

The vehicle control system is composed of longitudinal and lateral controllers as mentioned in Fig. 10.1. Lateral control for lane keeping (changing) while avoiding obstacles in designed/desired path, and longitudinal control for tracking (car following). In our work we will focus on lateral control designs. Where the complexity of the dynamic model of the vehicle, the environment, and driving situations make the transposition of the proposed solutions of path following problem into the context of mobile robots' field quite difficult.

Lateral control works on adjusting the steering angle such that the vehicle follows the desired reference path. It is considered as the most challenging part in vehicles dynamics researches while considering the non-linearity of the system, uncertainty and existing disturbances in the environment. Therefore, designing robust control

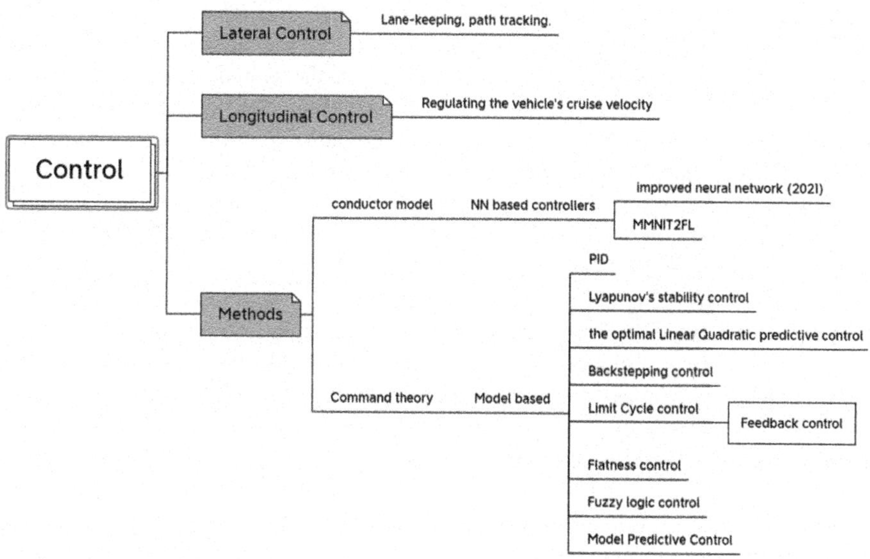

Fig. 10.1 Different lateral control approaches

laws capable of taking these requirements into account remains one of the main challenges today in vehicle dynamics. For ensuring such a maneuver the vehicle must have longitudinal control as CACC (Cooperative Adaptive Cruise Control), ACC (Adaptive Cruise Control) and CC (Cruise Control) to provide robust and string stable car-following in urban environments [3].

Lot of works on vehicle control are done using Optimal Linear Quadratic predictive control which provides optimal control [4], $H\infty$ controller-based approaches [5, 6], Sliding Mode Control for autonomous driving vehicle lateral maneuvering [7], also proposed in [8] and used with Backstepping control in [9], lane keeping control and lane changing control are concerned using Fuzzy logic control in [10, 11] Backstepping control [12, 13], Flatness control [14], Lyapunov's stability control [15], Model Predictive Control (MPC) [16], PID [17], or Limit Cycle control as authors in [18] used a modified the Limit Cycle Navigation method for the case of local sensing.

In this work we will focus on three methods. To begin, Model Predictive Control (MPC) is a method used in the trajectory tracking of autonomous vehicles, it deals with both of linear and nonlinear systems and which requires online optimization problem to be solved at each time t. The main disadvantage of this approach is the computation time of nonlinear MPC [16]. The MPC is designed for both of longitudinal and lateral vehicle control [19].

In addition, Proportional-Integral-Derivative PID is a control loop mechanism that employ feedback control which is widely used in industrial control systems. PID control is also a typical representative of classical control algorithms. As the PID controller is combined from three control terms, the P corrector brings speed to the system by reducing the rise time. It also reduces static error, but does not totally eliminate it. The integral action I (Ki) will deal with the static error and eliminate it. It therefore brings back precision, but degrades the transient response. Then, the derivative action D (Kd) improves the stability of the system, reduces overshoots and improves the transient regime. Thus, the performance of PID controllers in nonlinear systems is variable. Used also in lateral guidance as authors in [20] propose a new method to find operating points for multi-PID control. The PID controller can be modified and merged with neural networks to be able to achieve better results in controlling intelligent vehicles and provide safe driving which the biggest challenge in autonomous driving [17].

Last, Limit Cycle [21] is a control architecture that generates robot trajectories which are defined according to a set of differential equations where the stability is proven by Lyapunov function. This method generates cylindrical box of each object existing in the environment, where the center of the cylinder is respectively the object position. The robot follows the given cycle in order to avoid obstacle with an optimal path in case of a feasible collision-free while the safety distance between vehicles is respected. This approach is known by this optimality and it uses specific reactive rules which allows to the robot to avoid deadlocks, local minima and oscillations. Furthermore, it exists an extension of limit cycle called Parallel Elliptic Limit-Cycle 'PELC' [22], which is characterized by its elliptic bounding boxes instead cylindrical ones. As in [23] authors combine two methods to predict the robot trajectory to avoid obstacles. The approach uses the 'ELC' Elliptic Limit-Cycle which is a mathematical method and neural network method called the 'DBN' Dynamic Bayesian Network based which is maneuvering recognition based for decision-making. These three methods have better performance compared to other works. The reason why we intend to compare both of MPC and PID then with LC in future work.

10.3 Path Tracking

In this section we will discuss path tracking task for robot mobile. This task can be achieved while the automated robot is being connected to vehicle infrastructure communication (V2I) in order to obtain the precise vehicle location information. The main challenge is to be able to track path in real and complex scenarios where the given location information can be wrong [24]. The communication system between an autonomous driving and its surrounding road elements is indispensable [25], this communication can include vehicle to vehicle (V2V), vehicle to infrastructure (V2I), vehicle to pedestrian (V2P) and vehicle to network (V2N). Recently, researches are working on some methodologies which includes sensing schemes and detection algorithms utilizing a smartphone for drowsiness detection and abnormal driver behavior detection [26].

In this paper we propose a comparison between two methods already used in the literature separately or combined to overcome drawbacks of each. However, there are few works done where authors compare PID and MPC performances for mobile robot local path following as in [26, 27]. This work intends to achieve path following through two phases, starting by motion planning or defining the reference trajectory to be followed then adapt the controllers to control the vehicle's lateral movement.

10.3.1 Motion Planning

Motion planning or path planning by definition is a set of mathematical and computer techniques allowing to calculate trajectories for a kinematic system, with the constraint of the absence of collision. It is a very active field of research which already has a number noticed of applications especially in autonomous navigation. Motion planning is an essential module for automated vehicles, especially in scenarios of autonomous navigation. The existing path planning approaches that are based on optimization have been widely successful in autonomous driving.

10.4 The Control Problem of the Two Approaches

10.4.1 PID Controller

In autonomous driving, the vehicle's lateral movement is controlled by a lateral control algorithm. The PID regulator, or PID controllers (proportional, integral, derivative) is a series corrector which acts on both the gain and the phase margin, making it possible to improve the performance of a closed loop-controlled system. This is the most widely used method in industry in general. Also, the PID controllers are ideally suitable for first and second order equations. In this work we tried to model the vehicle's lateral control by carrying out a program on MATLAB using mathematical and geometric tools. In this case we will not consider external parameters such as disturbances caused by the wind, the variation in road grip.

The standard form of a PID control is written as:

$$u(t) = K_p \varepsilon(t) + K_i \int_{t_0}^{t_i} \varepsilon(t) d\tau + K_d \frac{d}{dt} \varepsilon(t) \qquad (10.1)$$

The controller acts on the error $\varepsilon(t)$ in three ways: Proportional action which consists in multiplying $\varepsilon(t)$ by a gain K_p. Integral action where the error $\varepsilon(t)$ is integrated and multiply by the gain K_i. Then, The Derivative action where the error $\varepsilon(t)$ is derived and multiply by the gain K_d. There are many adjustment processes for the gains K_p, $K_i et K_d$ in order to obtain a fast, precise and a stable system.

Before modeling the trajectory to be followed by the vehicle, we consider the vehicle as a material point. So that the variation in the steering angle of the vehicle is the rate of variation of the trajectory, which means that we vary the slope of the trajectory at each space step.

10.4.2 MPC Controller

The most known challenging control design problems in this field include non-linear vehicle dynamics, fast sampling time and limited computing resources on automated hardware. The MPC controller (Fig. 10.2) has the ability to systematically consider nonlinearity of the system, future predictions and operating constraints of the control system framework [28].

In this subsection, Model Predictive Controller (MPC) technique based on dynamic model will be presented below. The proposed MPC is inspired from [29] where the authors developed a control for path following of autonomous vehicles and discuss designing tracking path controllers using MPC model based on scenario avoidance obstacle on the highway with several variations in speed. The MPC can be generally applied to nonlinear MIMO (Multiple Input–Multiple Output)

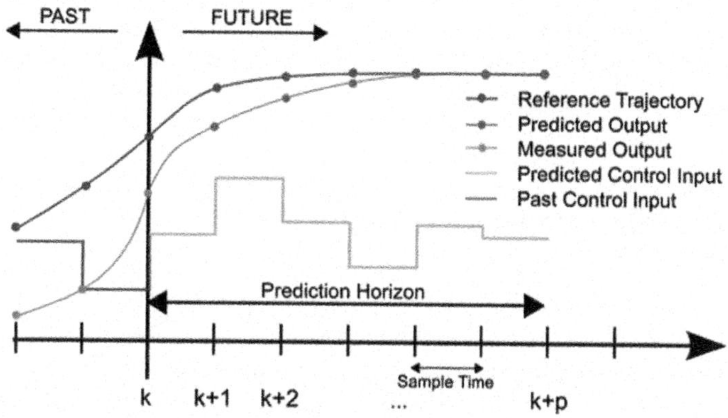

Fig. 10.2 MPC strategy for a reference trajectory tracking

systems, natural consideration of both states and control constraints, has approximately optimal control, widely used in industrial applications since the mid of 1970s and requires online optimization.

From input, output point of view, the vehicle model as a system can be simplified as:

$$\begin{bmatrix} \dot{x} \\ \dot{y} \\ \dot{\theta} \end{bmatrix} = \begin{bmatrix} vcos\theta \\ vsin\theta \\ \omega \end{bmatrix} = \begin{bmatrix} cos\theta & 0 \\ sin\theta & 0 \\ 0 & 1 \end{bmatrix} \begin{bmatrix} v \\ \omega \end{bmatrix} \tag{10.2}$$

where v is the linear velocity input and ω is the angular velocity input. The MPC strategy can be summarized in three main phases: prediction, online optimization and receding horizon implementation.

By defining the desired trajectory to follow X^{ref} and the control objectives, we can apply the MPC controller steps:

$$\text{Running costs}: \ell(x, u) = \| X_u - X^{ref} \|_{\mathcal{Q}}^2 + \| U - U^{ref} \|_{\mathcal{R}}^2 \tag{10.3}$$

$$\text{Optimal control problem}: minimize\ J_N(x_0, u) = \sum_{k=0}^{N-1} \ell(X_u(k), u(k)) \tag{10.4}$$

$$subject\ to: X_u(k+1) = f(X_u(k), u(k)),$$

$$X_u(0) = X_0,$$

where J_N is the cost function for the evaluation of the running costs along the whole prediction horizon and a cost function to be minimized for optimal control action$u(k)$,

X_u, X^{ref} are the states and the reference values at prediction instant k, respectively. Function f is the prediction model, X_0 is the current state measurement which will be updated online at every sampling instant.

10.5 Simulation and Experimental Results

Many researches assumed that their proposed control system for the PID control is able to achieve its objective even though it was sometimes less robust in maintaining its performance under multiple and various conditions as load change, cornering stiffness and different longitudinal speed. In the other hand, the control system for the MPC, its optimizer made sure that the predicted future trajectory of the vehicle output tracks the given reference trajectory; State of Art (SOTA) shows that MPC was more robust in maintaining its performance under same conditions mentioned before in the case of the PID control.

To set up the PID control, we give the vehicle an initial position at X(5,50). Various test cases were done trying to modify the three PID parameter in order to obtain the best reference trajectory following. Where increasing Ki parameter which controls the precision of the response eliminates the static error. So as to eliminate the oscillations caused by the increase in the gain Ki, we will decrease the proportional gain Kp which controls the speed of the response, and increase the derivative gain Kd which controls the stability. The simulation result of PID controller is shown in Fig. 10.3.

In our study case, we adopt a simple vehicle model (10.2) then we assume a constant speed and the vehicle's initial position is set at X(5,5). Concerning the MPC simulation, we fixed the vehicle's velocity at 0.5. The Q and R are weighting matrices of appropriate dimensions for states and controls, respectively. The prediction horizons were set to N = 5. The gain Q value are 3×3 matrix diagonal 1, 0.98 and 0.5. Then the gain R value are 2×2 matrix diagonal 0.0005. The simulation time is set to 400 s and the sampling time is set to 0.5 s. The simulation result of MPC controller is shown in Fig. 10.4.

10.6 Analysis of PID and MPC Results

PID model has a good performance in our study case where we ignore external disturbances and lateral dynamics. In complex system, we notice that the PID controller is unable to correct system behaviour where the system has many disturbances. The PID is faster than MPC, the MPC model is known by its major disadvantage which is the computation time. But it has many advantages over other possible solutions like the uncomplicated formulation, based on well understood concepts, explicit handling

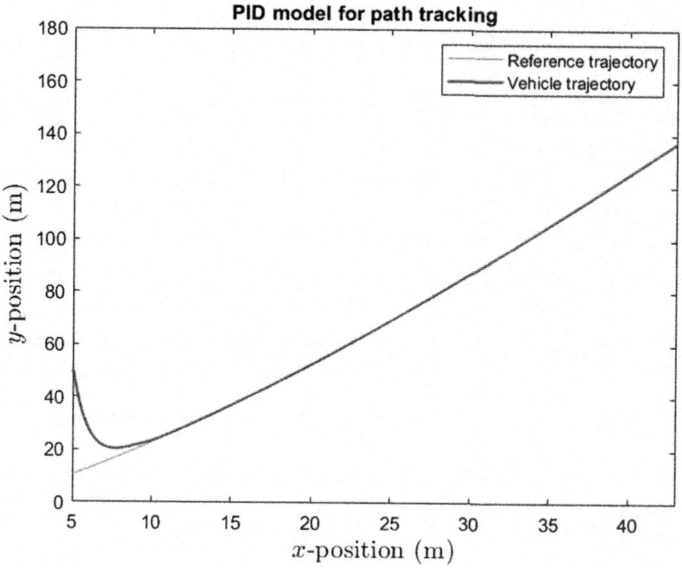

Fig. 10.3 PID control for reference trajectory following

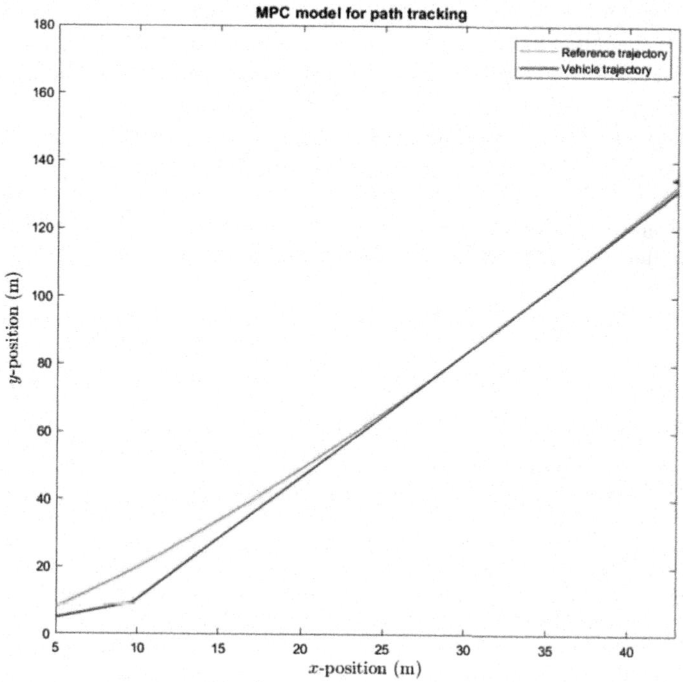

Fig. 10.4 MPC approach for reference trajectory following

of constraints explicit use of a model, well understood and defined tuning parameters and then the ability to predict future behaviour of the system and take it into consideration.

Model predictive control MPC can be used to control complex systems with multiple inputs and outputs MIMO where the simple PID controller is insufficient. This technique is particularly interesting when the systems have large delays, reverse responses and many disturbances. Some works intend to combine both of PID and MPC for better and faster performances and more stability and especially for fast response. Furthermore, it is suggested that a fixed control law for PID controllers is more used when a shorter time and higher velocities are sought, and as mentioned before high values for Ki and Kd factors are proposed. Moreover, MPC control is proposed as a function of the path to be followed because its flexibility improves the results.

10.7 Conclusion and Future Work

In this paper, various approaches used in the vehicle lateral control are introduced. Non-linear dynamics-based tracking controller (MPC) is suggested, which is capable of producing continuous and smooth velocity commands. The tracking controllers' simulations show that the system is asymptotically stable, and that the tracking errors are ensured to move toward to zeros. Also, we compared obtained the results.

As a future direction, we propose or evaluate a method called Limit Cycles for lateral dynamics. This method is rarely used in this field despite its advantages. This control gives an optimal path to follow while avoiding obstacles.

Acknowledgements This work is done with the support of the 'Région Bourgogne Franche-Comté' through the 'Itinéraire Chercheur Entrepreneur' funding grant.

References

1. Yang, B., Liang, M., Urtasun, R.: HDNET: exploiting HD maps for 3D object detection. In: Conference on Robot Learning, pp. 146–155. PMLR (2018, October)
2. Abdallaoui, S., Aglzim, E.H., Chaibet, A., Kribèche, A.: Thorough review analysis of safe control of autonomous vehicles: path planning and navigation techniques. Energies **15**(4), 1358 (2022)
3. Flores, C., Milanés, V.: Fractional-order-based ACC/CACC algorithm for improving string stability. Transp. Res. Part C: Emerg. Technol. **95**, 381–393 (2018)
4. Rui, W., Yi-Ming, S., Mei-Tong, L., Hao, Z.: Research on bus roll stability control based on LQR. In: 2015 International Conference on Intelligent Transportation, Big Data and Smart City, pp. 622–625. IEEE (2015, December)
5. Sun, P., Jing, Y.: Design of robust H controller for a class of uncertain systems. J.-Northeast. Univ. Nat. Sci. **27**(2), 119 (2006)

6. Huang, X., Zhang, H., Zhang, G., Wang, J.: Robust weighted gain scheduling H1 vehicle lateral motion control with considerations of steering system backlash-type hysteresis. IEEE Trans. Control Syst. Technol. **22**(5), 1740–1753 (2014)
7. Lee, S.H., Chung, C.C.: Predictive control with sliding mode for autonomous driving vehicle lateral maneuvering. In: 2017 American Control Conference (ACC), pp. 2998–3003. IEEE (2017, May)
8. Tagne, G., Talj, R., Charara, A.: Higher-order sliding mode control for lateral dynamics of autonomous vehicles, with experimental validation. In: 2013 IEEE Intelligent Vehicles Symposium (IV), pp. 678–683. IEEE (2013, June)
9. Kang, C.M., Kim, W., Lee, S.H., Chung, C.C.: Backstepping control method with sliding mode observer for autonomous lane keeping system. IFAC-PapersOnLine **50**(1), 6989–6995 (2017)
10. Jin, Z.H.A.O., Lefranc, G., El Kamel, A.: Lateral control of autonomous vehicles using multi-model and fuzzy approaches. IFAC Proc. Vol. **43**(8), 514–520 (2010)
11. Taghavifar, H., Rakheja, S.: Path-tracking of autonomous vehicles using a novel adaptive robust exponential-like-sliding-mode fuzzy type-2 neural network controller. Mech. Syst. Signal Process. **130**, 41–55 (2019)
12. Jiang, J., Astolfi, A.: Lateral control of an autonomous vehicle. IEEE Trans. Intell. Veh. **3**(2), 228–237 (2018)
13. Pang, H., Yao, R., Wang, P., Xu, Z.: Adaptive backstepping robust tracking control for stabilizing lateral dynamics of electric vehicles with uncertain parameters and external disturbances. Control Eng. Pract. **110**, 104781 (2021)
14. Menhour, L., d'Andréa Novel, B., Boussard, C., Fliess, M., Mounier, H.: Algebraic nonlinear estimation and flatness-based lateral/longitudinal control for automotive vehicles. In: International Conference on Intelligent Transportation Systems (2011)
15. Elsayed, M., Hammad, A., Hafez, A., Mansour, H.: Real time trajectory tracking controller based on Lyapunov function for mobile robot. Int. J. Comput. Appl. **975**, 8887 (2017)
16. Falcone, P., Borrelli, F., Asgari, J., Tseng, H.E., Hrovat, D.: A model predictive control approach for combined braking and steering in autonomous vehicles. In: 2007 Mediterranean Conference on Control & Automation, pp. 1–6. IEEE (2007, June)
17. Han, G., Fu, W., Wang, W., Wu, Z.: The lateral tracking control for the intelligent vehicle based on adaptive PID neural network. Sensors **17**(6), 1244 (2017)
18. Grech, R., Fabri, S.G.: Trajectory tracking in the presence of obstacles using the limit cycle navigation method. In: Proceedings of the 2005 IEEE International Symposium on, Mediterrean Conference on Control and Automation Intelligent Control, 2005, pp. 101–106. IEEE (2005, June)
19. Attia, R., Orjuela, R., Basset, M.: Combined longitudinal and lateral control for automated vehicle guidance. Veh. Syst. Dyn. **52**(2), 261–279 (2014)
20. Monot, N., Moreau, X., Benine-Neto, A., Rizzo, A., Aioun, F.: New method to find operating points for multi-PID control. Application to vehicle lateral guidance. IFAC-PapersOnLine **51**(4), 172–177 (2018)
21. Adouane, L.: Orbital obstacle avoidance algorithm for reliable and on-line mobile robot navigation. In: 9th Conference on Autonomous Robot Systems and Competitions (2009)
22. Adouane, L.: Reactive versus cognitive vehicle navigation based on optimal local and global PELC∗. Robot. Auton. Syst. **88**, 51–70 (2017)
23. Iberraken, D., Adouanc, L., Denis, D.: Multi-controller architecture for reliable autonomous vehicle navigation: combination of model-driven and data-driven formalization. In: 2019 IEEE Intelligent Vehicles Symposium (IV), pp. 245–251. IEEE (2019)
24. Ge, J., Pei, H., Yao, D., Zhang, Y.: A robust path tracking algorithm for connected and automated vehicles under i-VICS. Transp. Res. Interdiscip. Perspect. **9**, 100314 (2021)
25. Chan, T.K., Chin, C.S.: Review of autonomous intelligent vehicles for urban driving and parking. Electronics **10**(9), 1021 (2021)
26. Chan, T.K., Chin, C.S., Chen, H., Zhong, X.: A Comprehensive review of driver behavior analysis utilizing smartphones. IEEE Trans. Intell. Transp. Syst. **21**(10), 4444–4475 (2020)

27. Pacheco, L., Luo, N.: Testing PID and MPC performance for mobile robot local path-following. Int. J. Adv. Rob. Syst. **12**(11), 155 (2015)
28. Leman, Z.A., Ariff, M.H.M., Zamzuri, H., Rahman, M.A.A., Mazlan, S.A.: Model predictive controller for path tracking and obstacle avoidance manoeuvre on autonomous vehicle. In: 2019 12th Asian Control Conference (ASCC), pp. 1271–1276). IEEE (2019, June)
29. Daoud, M.A., Osman, M., Mehrez, M.W., Melek, W.W.: Path-following and adjustable driving behavior of autonomous vehicles using dual-objective nonlinear MPC. In: 2019 IEEE International Conference on Vehicular Electronics and Safety (ICVES), pp. 1–6. IEEE (2019, September)

Chapter 11
Dark-Matter Search Optimiser

P. A. Salgado and T.-P. Azevedo Perdicoúlis

Abstract A version of the Gravitational Search Algorithm that takes advantage of the repulsive forces of dark matter to explore the space is presented in this article. The classification of matter into dark and gravitational particles is used to balance between exploration and exploitation of the feasible set of solutions, respectively. The idea is to overcome some problems of the Gravitational Search Algorithm as, for instance, trapping into local optima and the choice of parameters. Three different problems are used to demonstrate the potential of the algorithm when compared with the Particle Swarm Optimiser. In the future, we plan to do a thorough assessment of the algorithm on benchmark problems as well as on some applications.

Keywords Gravitational search algorithm · Global optimisation · Dark-matter · Intelligent search algorithms

11.1 Introduction

In classical and standard terms, optimisation consists in maximising/minimising a real function by systematically choosing input values from within an allowed set and computing the value of the function. The generalisation of optimisation theory and techniques to other disciplines constitutes a large area of applied mathematics. A variety of different types of objective functions and different types of domains can be considered [1]. Besides, in stochastic optimisation, the uncertainty is incorporated into the model [2]. In this case, the goal is to find a solution that is feasible for all data and optimal in some sense. Opposing to the classical approach, heuristic methods designate a computational procedure that determines an optimal solution by iteratively trying to improve a candidate solution with regard to a given measure of quality.

P. A. Salgado
CITAB and Eng. Dep. at ECT CITAB, UTAD, 5000-801 Vila Real, Portugal

T.-P. Azevedo Perdicoúlis (✉)
UTAD and ISR-Coimbra, 5000-801 Vila Real, Portugal
e-mail: tazevedo@utad.pt

M. A. Laribi et al. (eds.), *Advances in Automation, Mechanical and Design Engineering*, Mechanisms and Machine Science 121,
https://doi.org/10.1007/978-3-031-09909-0_11

Heuristics make few or no assumptions about the problem being optimised and can search large spaces of candidate solutions towards finding sub-optimal solutions at a reasonable computational cost without being able to guarantee either feasibility or optimality, or even in many cases to state how close to optimality a particular feasible solution is. Heuristics implement some form of stochastic search optimisation like evolution programming, genetic algorithms, genetic programming, and differential evolution [3, 4]. Other methods having a similar meaning as heuristic are derivative-free [5], direct search [6], and black-box optimisation techniques [7]. Intelligent search methods are often preferred to solve nonlinear problems which use intensive mathematical equations, due to the intricate mathematics used in classical methods. The major advantage of the heuristic approach, comparing to analytical, is its simplicity, although it sacrifices accuracy and precision. A meta-heuristic is an iterative procedure to find near-optimal solutions in a more efficient way. The objective of meta-heuristics is to enlarge the aptitudes of heuristics by joining one or more heuristic methods. Popular approaches are Genetic Algorithm (GA) [8], Simulated Annealing [9], Tabu Search [10, 11], Particle Swarm Optimisation (PSO), Ant Colony Optimisation (ACO) [12], and Harmony Search [13]. However, analytical approaches are still being used in recent research given their advantage in explaining the physics and mechanisms behind the mathematical models. They are also suitable when it comes to the validation of numerical methods.

Gravitational search algorithms (GSA) are meta-heuristics used to calculate the global optimum of any type of function based on the Newton universal gravity and motion laws. It has first been presented in 2009 by Rashedi and Saryazdi [14], and during the last decade different versions have been proposed. This is one of the most recent techniques of evolutionary algorithms—a swarm-based search algorithm. PSO are global optimisation algorithms for dealing with problems in which a best solution can be represented as a point or surface in an n-dimensional space [15]. This type of search algorithm has been first proposed by Kennedy and Eberhart in 1995 [16, 17]. In [18] intelligent search methods are reviewed and compared and PSO is said to stand out in terms of its low number of parameters, simplicity of implementation, fast convergence and short computational times or even robustness. However, especially with complex problems, it may get trapped into a local maxima. Alike other swarm-based algorithms, in GSA, particles begin to look for the most appropriate solution by selecting a random location in the solution space. Then, they start off a journey towards the optimal solution, taking advantage of both the personal and the swarm's experiences.

In [19], a binary version of GSA was developed and its performance confirmed by solving different nonlinear benchmark problems. In [20] research was conducted on diverse variations of multi-objective PSO algorithms, since each of which might have its own strengths and weaknesses. Due to the high complexity of multi-objective problems, the efficiency of these methods has become a matter of concern. In this paper a new multi-objective meta-heuristic algorithm based on gravitational forces is proposed and applied to different test benches. The acquired results proved the superiority of the algorithm comparing with other pioneering techniques such as the MOPSO and a new multi-objective GSA. In [21] is proposed an improved

PSO-GSA algorithm for solving two kinds of classifications: surface water quality and the moving direction of robots. In [22], a new updating mechanism has been proposed to prevent premature convergence and stagnation in evolution. In [23], a new version of GSA was proposed and experiments were done on two sets of benchmark categories, namely standard functions and the problems of IEEE Congress on Evolutionary Computation (CEC) in 2013—this includes problems belonging to different categories such as unimodal, multi-modal, and unconstrained optimisation functions. In [24, 25] early optimisation works based on GSA are presented. Based on the literature, GSA is capable of providing more accurate, effective and robust high-quality solutions for most of the optimisation problems, such as in power system, controller design, network routing, sensor networks or software design. It has also been adapted to optimise the design of controllers, software, antennas and micro-grids. GSA has shown better performance in solving the optimisation problems compared to other previous algorithms such as PSO or ACO. A new meta-heuristic technique is proposed in [26], i.e., a GSA for optimisation of test paths. The results obtained from newly generated meta-heuristic algorithm are compared with Simple Genetic Algorithm (SGA) and show better performance. In [27], the phenomenon of premature convergence to a local maximum is addressed, since the particle diversity declines during the optimisation process and the PSO information is not shared. In addition, some interesting results have been obtained with the application of GSA in different fields, namely in the standard cell placement problem [28]. A hybrid algorithm was derived from both GA and GSA for feature set selection in [29]. In [30] the GSA algorithm is used to determine the optimal settings of the control variables of an optimal power flow (OPF) problem. In [31] GSA was employed to optimise the total annual cost of hybrid wind and solar renewable energy systems and having to satisfy the pre-designed load. Furthermore, GSA variants have been applied in solving many other problems such as the symmetric traveling salesman problem [32], the flow shop scheduling problem [33], feature selection [34], image enhancement [35], DNA sequence design problem [36] and to optimise a filter modelling parameters [37].

In this work, we developed a new algorithm also inspired in Physics. In order to improve the exploration capacity of the GSA algorithm, the particles are classified into light-matter and dark-matter, using a threshold value calculated from their current position. The repulsive forces between dark-matter particles (DP) and dark-matter and light-matter (GP) contribute to a broader exploration of the search space and may even untrap local minima. This new procedure is designated Dark-matter Search Optimiser (DSO). The algorithm potential is demonstrated in three examples: (i) a multi-modal function with an obvious maximum and a crowd of local maxima all with the same performance value, (ii) a multi-modal function with a set of local maxima with slightly different performance values, (iii) a neuronal network that implements the XOR operation. Applicability of this type of algorithms is very broad as one can see from the review of past work presented in the beginning of this section on GSA. DSO performance has been compared with the canonical PSO and the results are promising. PSO always finds a unique "sub"-optimum while DSO determines a set of solutions.

Hereby, we present this new version of the algorithm. In Sect. 11.2 the simple PSO algorithm is described to make this article more self-contained and facilitate its reading. The section is highly based on [38]. In Sect. 11.3 the Newton laws of gravity and movement are revisited with the new DSO algorithm being explained in Sect. 11.3.1. In Sect. 11.4 are described the problems used to assess DSO algorithm and its respective obtained results. Some final conclusions and directions for future work are stated in Sect. 11.5.

11.2 Particle Swarm Optimiser Algorithm

PSO algorithm is inspired on the dynamics of a bee swarm or a flock of birds flying over a place, aiming to find a point to land. The definition of a suitable point for the whole flock to land is a complex problem, since it depends on several issues, as for instance the maximisation of available food and minimisation of the risk of existence of predators. The movement of the birds can be understood as a choreography: the birds synchronically move for a certain period until the best place to land is defined and, then, the whole flock lands at once.

In the above description, the movement of the flock can only happen as described because all flock members are able to share information among themselves. Otherwise, most likely, each animal would land at a different point and at a different instant of time. Studies regarding social behaviour in the early 1990s stated that all birds of the flock searching for a good point to land are able to know the best point once it has been found by one of the members. In this way, each member of the flock balances between its individual and the flock knowledge experience—known as social knowledge. The criteria to assess whether a point is good or not, in this case, are the survival conditions found at a possible landing point. The effort to find the best point to land describes an optimisation problem: The flock must identify the best point—for example, the latitude and the longitude—in order to maximise the survival conditions of its members. To do so, each bird flies searching for and assessing different points, using several surviving criteria at the same time. Everyone and each of them has the advantage to know where its best location point is found until the global best for the whole flock is known. The goal of an optimisation problem is to determine a vector $x = (x_1, x_2, \ldots, x_d)$ that minimises/maximises a function $f(x)$ in a search space of dimension d. Variable $x \in \mathbb{R}^d$ is known as the position—in the flock case, e.g. the latitude and the longitude. Function $f(x)$ is called the fitness function or objective function and is used to assess how good a position x is. Consider a swarm of n particles P_i. For every one of these particles, we define at every iteration-t a position xi and a velocity vi that are updated in the following manner:

$$v^i(t+1) = wv^i(t) + c_1 r_1 (p^i_{best}(k) - x^i(t)) + c_2 r_2 (g^i_{best}(t) - x^i(t)) \quad (11.1)$$

$$x^i(t+1) = x^i(t) + v^i(t+1)\underbrace{T}_{=1}, \tag{11.2}$$

where w is an inertia constant weight, which is a positive value for the classical PSO version. This is an important parameter to balance between global search—also known as exploration—and local search—known as exploitation. Higher values are set for the first case and lower for the later. The main differences between the classical version of PSO and other versions lay in the definition of this factor. The first term of Eq. (11.1) denotes the contribution of the particles' previous motion into the current instant. For example, if $w = 1$, it means that the particle's current motion is fully influenced by its previous, that is, the particles keeps going in the same direction. On other hand, when $0 \leq w < 1$ it means that the particles may go in other directions and, therefore, it is likely to correspond to a broader exploration of the search space. However, if the parameter is chosen too small, the convergence of the iterative procedure may become too slow. The second term represents the individual cognition. It is calculated in terms of the difference between the particle's own best position, $p^i_{best}(t)$, and its current one $x^i(t)$. This term increases with the distance of particle P_i to its best position, therefore attracting the particle to its best position. c_1 is the individual-cognition parameter, a positive constant that weighs the importance of the particle's own previous experiences. r_1 is a random value within the interval [0, 1] that plays an important role to avoid premature convergence, and increasing in this way the likeness to find the global optima. The third term represents social learning. It works very alike the second one, where $g^i_{best}(t)$ represents flock's best position. Particles share information of the best point through this term.

Algorithm

1. Initialisation

    1.1. Initialise randomly x^i and v^i, $i = 1, \ldots, n$.
    1.2. Evaluate $f(x^i)$, $\forall i$.
    1.3. Initialise p^i_{best} with a copy of x^i, $\forall i$.

2. Iteration$-t$ to repeat until a stop criterium is satisfied. For every particle P_i, $i = 1, \ldots, n$

    i. Update x^i and v^i using Eqs. (11.1) and (11.2).
    ii. Evaluate $f(x^i(t))$.
    iii. $p^i_{best} \leftarrow x^i(t) sef(p^i_{best}) < f(x^i)$
    iv. $g^i_{best} \leftarrow x^i(t) sef(g^i_{best}) < f(x^i)$

11.3 Principles of Gravity and Movement of Particles

In this section, we review the newtonian principles that rule the movement of particles and the gravity laws. As is well known, the universe is composed of gravity matter and dark-matter. Therefore, two different types of particles exist: gravitational particles (GP), also known as gravitons, and dark-matter particles (DP). Gravitons have the capacity to reflect the light and its mass is, thus, observable while DP are not observable. In the universe, exists more DP than GP. GP attract each other by a force directly proportional to the product of its masses and inversely proportional to the distance to each other. Thus, considering two GP, P_i and P_j, with mass m_i an and m_j, respectively, particle P_i is subject to the attraction force exerted by particle P_j,

F_{ij}, and the acceleration, a_{ij}. Hence:

$$F_{ij} = G\frac{m_i(x^i)m_j(x^i)}{d(x^i, x^j)^2}\hat{u}_{ij}, i \neq j \tag{11.3}$$

$$a_{ij} = G\frac{m_j(x^j)}{d(x^i, x^j)^2}\hat{u}_{ij}, \tag{11.4}$$

where G is the attraction gravity constant. Both, force and acceleration, have direction $\hat{u}_{ij}$, i.e., from particle P_i to particle P_j. $d(x^i, x^j)$ is some distance definition, for instance the euclidean distance. Therefore, the larger the masses of two gravitons, the stronger is the attraction force between them. This causes a movement towards the particles of larger masses translated into having that particles of higher performance attract particles of higher performance.

On the other side, DP repel each other by a force directly proportional to the product of its masses and inversely proportional to the distance to each other. Thus, considering two DP P_i and P_j, with mass m_i and m_j, respectively, particle P_i is subject to the repulsion force exerted by particle P_j, F_{ij}, and the acceleration a_{ij}. Hence:

$$F_{ij} = -G_r\frac{m_i(x^i)m_j(x^j)}{d(x^i, x^j)^2}\hat{u}_{ij}, i \neq j \tag{11.5}$$

$$a_{ij} = -G_r\frac{m_j(x^j)}{d(x^i, x^j)^2}\hat{u}_{ij}, \tag{11.6}$$

where G_r is the repulsion gravity constant, $G = 10 \times G_r$. Likewise, $\hat{u}_{ij}$ defines the direction of F_{ij} and a_{ij}, and the larger the masses of two DP, also the stronger is the repulsive force between them. This causes a movement away from the particles of larger masses.

The force and acceleration between a GP, P_i, and a DP, P_j, and vice-versa, is defined in the following manner:

$$F_{ij} = -G_r \frac{m_i(x^i)m_j(x^j)}{d(x^i, x^j)^2} \hat{u}_{ij}, i \neq j \tag{11.7}$$

$$a_{ij} = -G_r \frac{m_j}{d(x^i, x^j)^2} \hat{u}_{ij} \tag{11.8}$$

The mass of a particle is not constant as it moves along the search space. In this work, it is represented by a performance function $f(x^i)$ that is defined in a search space X of dimension$-d$:

$$f : X \subset B_1 \times B_2 \times \cdots \times B_d \to \mathbb{R}_0^+,$$

since every particle P_i has a position $x^i = (x_1, \ldots, x_d), x_k \in B_k, k = 1, \ldots, d$, in the d$-$dimension space and $x_k^i = \left(x_{k1}^i, \ldots, x_{kk_1}^i\right) \in B_k$.

We consider a total of n particles, $\{P_i\}_{i=1,\ldots,n}$ that move in the search d$-$space. The domain X of function f represents the set of positions that correspond to the reachable set for the mass values in the search space.

For every particle, at every position x^i, its mass is given by the performance function $f(x^i) = m_i, i = 1, \ldots, n$. (We may consider a normalised $f(x)$, i.e., $f(x^i) \in [0, 1]$.)

The total force acting on particle P_i is given by:

$$F^i = \sum_{j \neq i}^{K} F_{ij}(x^i, x^j), i \neq j, \tag{11.9}$$

as we may select the K particles P_j with stronger $F_{ij}(x^i, x^j)$. Thus,

$$a^i = \frac{F^i}{m_i}. \tag{11.10}$$

Furthermore, the position and velocity of particle P_i are updated along time, t, in the following way:

$$v^i(t+1) = v^i(t) * \alpha + a^i(t) \underbrace{T}_{1}, \tag{11.11}$$

$$x^i(t+1) = x^i(t) + v^i(t+1)T, \tag{11.12}$$

where α is a tuning factor, chosen randomly in [0, 1].

11.3.1 Dark-Matter Search Optimiser

The algorithm described herein determines which particles have the largest mass as they move around the search space, and finds a unique collective optimum. It is a modification of the GSA. As mass is represented $\mathrm{f(x), x \in X}$, since we want to maximise the particles' mass, we calculate the maximum of $\mathrm{f(x^i), \forall i = 1, \ldots, n, x^i \in X}$, along time, that is done taking advantage of the cleavage between GP and DP. Particles of high performance exert stronger repulsive forces among each other, promoting a more efficient exploration of the search space and freeing trapped local optima. On the other side, the repulsive force between DP and gravitons pushes the latter away, contributing also to faster convergence to the GP that hold the larger masses. Also, attraction forces between high performance gravitons enhance the global optimum search.

DSO algorithm description

Step–1: Definition of search space/performance function;

Step–2: Random initialisation of particle movement;

Step–3: Assessment of every particle performance;

Step–4: Definition of the threshold δ;

Step–5: Classification of every particle into either GP or DP using a threshold δ;

Step–6: Calculation of every particle force and acceleration according to its classification, done in Step 5.

Step–7: Update particles' position and velocity;

Step–8: Repeat Steps 3–7 until a stop criterion is satisfied.

Next, we detail the initialisation of the whole procedure and also the steps of every iteration. In the end of this subsection, we explain why we need to deal with local solutions.

Initialisation

1. Define performance function, f, and search space, X.
2. Generate random initial positions, $x^i(0), i = 1, \ldots, n, inX$. As $B_k = [a_k, b_k], x_k^i \sim U[a_k, b_k], k = 1, \ldots, d$.
3. Generate random initial velocities in X. These should be values close to zero to make sure that the particles are not "spat" out from the search space: $v^i(0) \approx 0, \forall i$.
4. Initialise as zero the different accelerations, for every i:
    i. between two GP: $a_{gg}(0)$
    ii. between two BP: $a_{dd}(0)$
    iii. between a GP and a DP:$a_{gd}(0)$
    iv. between a DP and a GP: $a_{gd}(0)$
5. Initialise as zero potential energy associated to $P_i, \forall i$
    i. Initial gravitational energy: two GP: $U_{gg}(0)$
    ii. Repulsive force energy: two DP: $U_{dd}(0)$
    iii. Repulsive force energy: GP and DP: $U_{gd}(0)$

iv. Repulsive force energy: DP and GP: $U_{dg}(0)$
v. Save in $U^i = [U_{gg}, U_{dg}, U_{gd}, U_{dd}], \forall i$

6. Calculate the initial mass of every particle using the performance function: $f(x^i(0)), \forall i and x^i(0) \in X$.
7. Based on the particle initial position, define a threshold δ that decomposes the whole set of particles: (i) gravitational matter above the threshold; (ii) dark-matter below the threshold. To do this:

   1. Define β as the ratio of dark-matter; (*Sug:* > 0.5 since there is more dark-matter than gravitational)
   2. Define threshold δ:
      i. determine the particle with the largest mass: best$M(x^i(0))$; Initialise Mbest with this value;
      ii. calculate the mean value of the mass for all the particles: mean$M(x^i(0))$
      iii. Calculate $\delta = \mathrm{meanM}(x^i(0)) + \beta(\mathrm{bestM}(x^i(0)) - \mathrm{meanM}(x^i(0)))$; (A different δ can be defined for $\neq$ regions of X)
   3. Save GP's indices in I and DP's indices in I_r;

8. Initialise local mass centres for the gravitational matter.
9. Create a data structure to store, for every iteration: GP mass mean value, DP mass mean value, threshold δ.
10. Define all necessary parameters.

Iteration-*t*

1. To every $P_i, i = 1, \ldots n$, is associated a position x^i, a velocity v^i and a mass m_i:

   1.1. Calculate $\hat{u}_{ij} = x^i - x^j$ and $d_{ij} = \|x^i - x^j\|_2$;
   1.2. Define some noise $r_{n_i} \sim U[\frac{1}{2}, 1], i = 1, \ldots, n$;
   1.3. For every GP$-i \in I$
      i. Calculate potential energy: $U_{gg} = -G \sum_{j \in I} \frac{m(x^j)}{d_{ij}}, j \neq i$
      ii. Determine the particle with maximum mass: $m_j = \underset{i \in I}{max} m(x^i), j \neq i$
      iii. Determine the maximum acceleration particle: $a_{gg} = r_{n_j} G m_j \hat{u}(i, j)$

      (r_{n_j} is added to simulate an imperfect system)

      iv. To improve algorithm convergence, formula (11.9) is modified as: $F_i = \underset{j \neq i, j \in I}{max} F_{ij}(x^i, x^j)$
      v. Calculate the repulsive forces resultant on GP$-i$—i.e., when $m(x^i)$ interacts with every mass $m(x^j), j \in I_r : F^i = \sum_{j \in I_r} F_{ij}(x^i, x^j)$,

and its respective potential energy and acceleration: $U_{gd} = G_r \sum_{j\in I_r} \frac{m(x^j)}{d_{ij}}$ and $a_{gd} = -G_r \sum_{j\in I_r} \frac{m(x^j)}{d_{ij}} \hat{u}(i, j)$.

1.4. For every DP$-i \in I_r$

i. Calculate potential energy and acceleration: $U_{dd} = -G_r \sum_{j\in I_r} \frac{m(x^j)}{d_{ij}}$, $j \neq i$ and

$$a_{dg} = -G_r \sum_{j\in I} \frac{m(x^j)\hat{u}(i, j)}{d_{ij}}$$

ii. Determine other repulsive forces, potential energy and accelerations on this particle:

$$U_{dg} = G_r \sum_{j\in I} \frac{m(x^j)}{d_{ij}} \tag{11.13}$$

$$a_{dd} = -G_r \sum_{j\in I_r} \frac{m(x^j)\hat{u}(i, j)}{d_{ij}} \tag{11.14}$$

1.5. Adjust velocity taking into account all different accelerations for every particle i, a^i: $v^i(t+1) = \alpha v^i(t) + a^i(t)$, $i = 1, \ldots, n$, where α is a damping factor (e.g. $\alpha = 0.95$) and $a(t) = a_{gg} + a_{gk} + a_{kg} + a_{kk}$.

1.6. Particle movement adjustment:

$$x^i(t+1) = x^i(t) + v^i(t+1), i = 1 \in I_r$$

2. Replacement of all the particles that went out of the search space back inside
3. Calculate the performance for every particle P_i, $\forall i$: M = performance(X), i.e., $X = \left(x^1, \ldots, x^n\right)$ and $M = \left(m(x^1), \ldots, m(x^n)\right)$.
4. Save the information of t-th iteration:

$$\text{Inform}(t) = [\text{mean}(M(I)), \text{mean}(M(Ir)), \delta]$$

$$U(t) = \text{sum}\left(\left[U_{gg}, U_{gd}, U_{dd}, U_{dg}\right]\right)$$

5. Check whether the solution has improved and save it if it is the case:

i. Determine the iteration optimum: $m_j = \max m(x^i)$

ii. Compare to see whether is better than the current optimum: $m_j >$ Mbest?

If yes:

-Update the best solution: Mbest $= m_j$;

-Update the best position: $Pbest = xj$;

6. Recalculate δ and redo the classification of all the particles into GP and DP; update

$$I \text{ and } I_r.$$

7. Update the mass centers for the gravitational matter: these are the best local solutions (see Sect. 11.3.1).

Local solutions Next, a procedure to find local mass centers is described. This algorithm finds groups centred in regions of higher mass-value—either a pre-established group/point. In every iteration-t, it is applied solely to gravitational-matter.

Algorithm description

1. Verification of groups existence.
   i. If l groups already exist, find out which particles are not yet classified (*i.e.*, $K = 1$).
   ii. If no groups yet, all the particles are not classified.
2. Sort all the mass-values in ascending order.
3. Get the first non-classified one (because it detains the highest value): P_g of coordinates x^g.
4. Define
   i. a region of radius r and center x^g
   ii. Mcentro $= m(x^g)$
5. Determine which particles belong to the same group. To belong to the same group, a particle-j has to verify the following criteria:

   -belongs to the region defined in Step 4
   -$m(x^j) > \alpha *$ Mcentro where α is a pre-defined parameter

6. Set all the particles eligible to enter the group as classified (*i.e.*, $K = 0$).
7. Go through all the mass-values and get the largest one amongst the non-classified ones.
8. Repeat Steps 3–7 until all particles have been classified.

The result of this local algorithm is a set of groups characterised by the coordinates of its center, and its mass, defined as $\max m_j$, where m_j are the masses of all the particles that belong to the group. For every particle, a vector J gives the information about the group to which the particle pertains.

11.4 Numerical Examples

The viability of DSO is demonstrated in three different problems: two multi-modal functions, with completely different shapes, and an xor neural network. In Example 1, the function of has a notorious global maximum while the one of Example 2 has many local maxima which are slightly different. For every problem the results were compared with the ones obtained with a simple PSO. These are promising, since the DSO always manages to find an approximation to the global optimum for all the problems, although it may obtain various solutions while the PSO always presents a unique solution. The DSO algorithm works better with a high number of particles, since it fosters a more intensive exploitation/ exploration of the feasible space.

11.4.1 Problem 1

Consider $f(x) = \frac{\sin x}{x}$, $x \in [-1, 1] \times [-1, 1]$, represented in Fig. 11.1, that has an optimum and a crowd of sub-optimal solutions. DSO has been applied to this problem with 100 particles (≈20 GP) and 400 iterations.

Figure 11.2 shows the evolution of the classification of particles into light-matter and dark-matter using threshold δ: DP diminishes along iterations. As potential energy is understood as the work necessary to drive the mass to infinity, an increase on the number of GP leads to an increase of potential energy. There are more particles eligible to the best particle and contributing to the search of the interesting zones of the feasible space.

In Fig. 11.3 the performance of the two algorithms is compared. PSO determines a unique solution while DSO may devolve a set of solutions (see Fig. 11.3b), one on the peak and a set of local solutions on the crowd, although in the end it can tell where

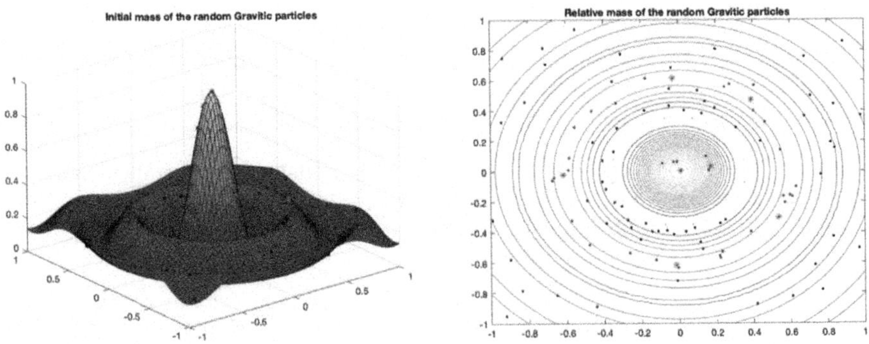

Fig. 11.1 Left: representation of GP (red) and DP (black) on the surface; right: representation of the particles' mass on the level-curves

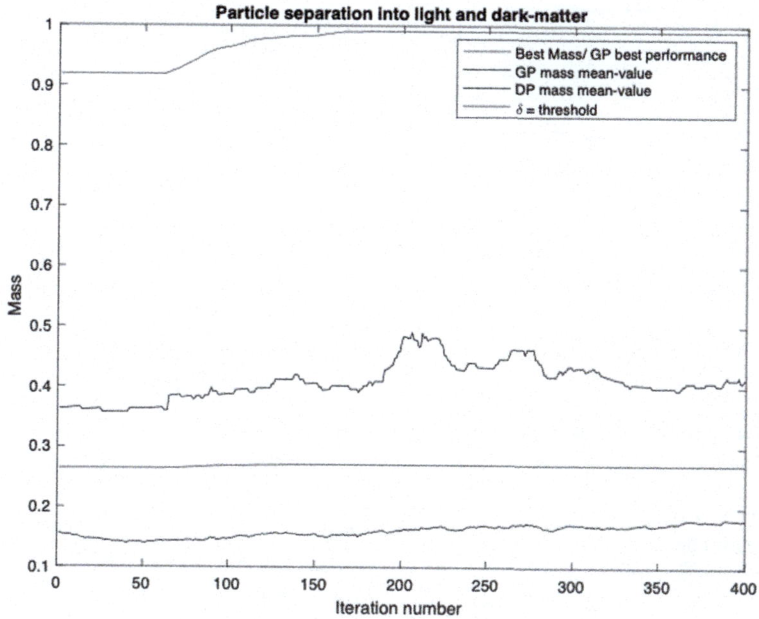

Fig. 11.2 Threshold definition and respective GP/DP quantity along iterations

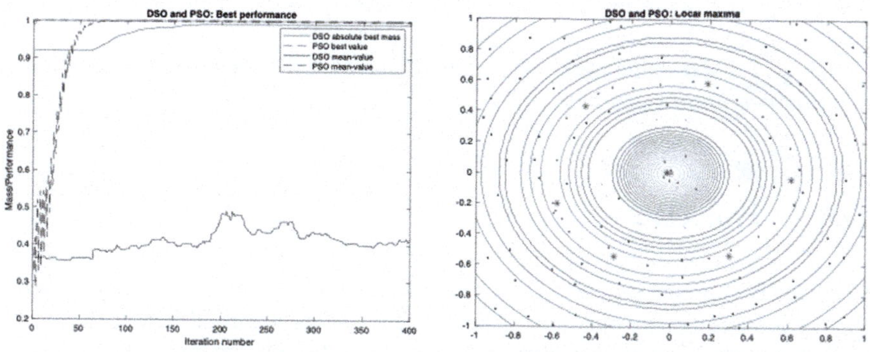

(a) Algorithms convergence.

(b) DSO solutions (red)/ PSO solutions (green).

Fig. 11.3 Algorithms' convergence and solution set

the absolute maximum lays. However, it may happen that DSO also only devolves a unique solution: the absolute maximum.

The features of group solutions described in Sect. 11.3.1 emulate local PSO solutions. In Fig. 11.4 one can see the evolution of capturing the local solutions by the DSO algorithm as the search space is being explored. The number of groups may change along the iterative procedure as it can be observed in the figure on the left. The figure on the right shows the evolution of the groups (in different colours)

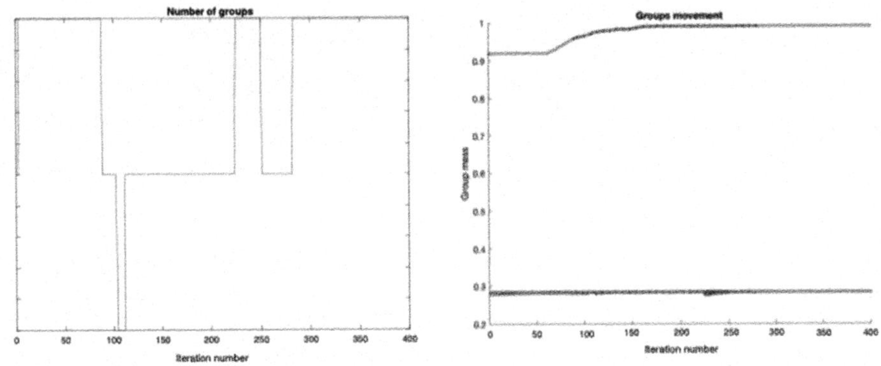

(a) Number of groups: DSO.

(b) Groups' evolution.

Fig. 11.4 Number of local groups along iterations

Table 11.1 Example 1: Algorithms comparison

	Solution: P_i	Performance
DSO	(−0.020, 0.005)	0.991
PSO	(0.000, 0.000)	1.000

and everyone of those presents the flock-like behaviour of the PSO algorithm. It is interesting to observe, in this simulation, the clear search for the peak and for the crowd.

Table 11.1 shows the result of a simulation where the solution captured by both algorithms is almost the same.

11.4.2 Problem 2

Consider $f(x) = \sin x \cos x e^{a(x-x_0)}$, $x \in [-1, 1] \times [-1, 1]$ and $a \approx 0$, represented in Fig. 11.5, whose many maximising points have values slightly different.

There is also an evolution of particles along iterations using threshold δ, although the decrease of DP is more accentuated in this example. In Fig. 11.6 the performance of the two algorithms is compared.

The capture of local solutions in Fig. 11.7 by the DSO is also similar in this example: due to the characteristics of the problem, this number keeps almost constant along the whole process and being equal to the peaks in the final iteration: 10.

Table 11.2 shows the result of a simulation where the solution captured by both algorithms is almost the same. But there are situations where DSO-solution is frankly better than the one obtained by the PSO.

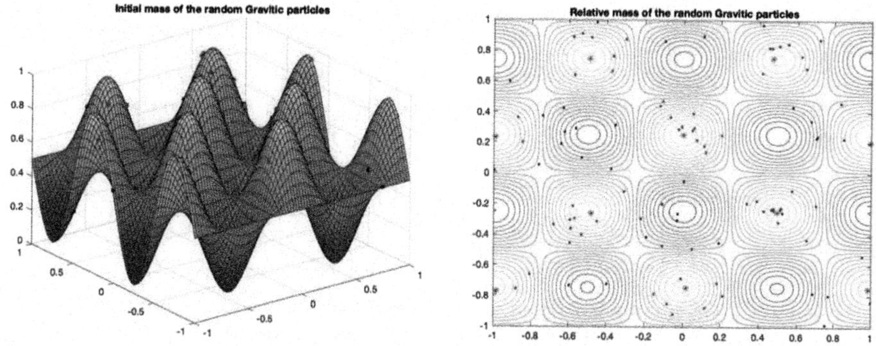

Fig. 11.5 Left: representation of GP (red) and DP (black) on the surface; right: representation of the particles' mass on the level-curves

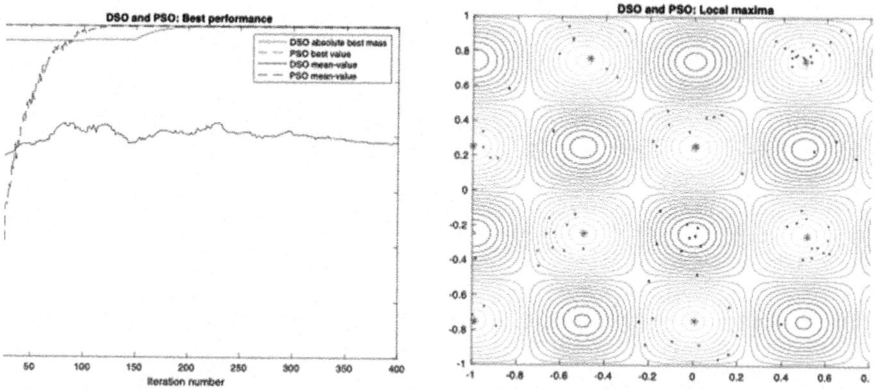

(a) Algorithms convergence.

(b) Local maxima: DSO (red) and PSO (g

Fig. 11.6 Algorithms' convergence and solution set

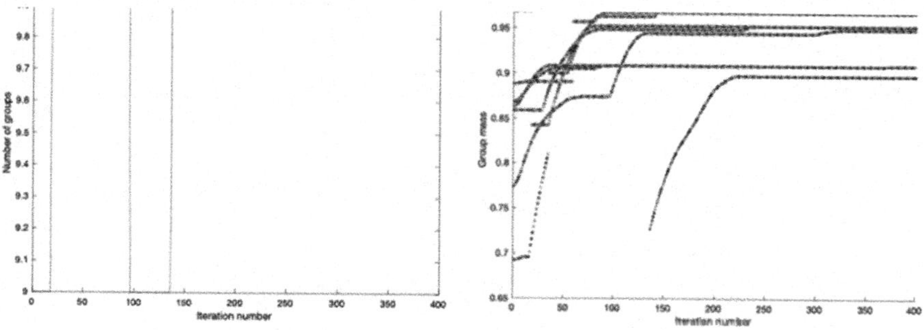

Fig. 11.7 Number of local groups along iterations

Table 11.2 Example 2: Algorithms comparison

	Solution: P_i	Performance
DSO	(0.006, 0.256)	0.999
PSO	(0.000, 0.250)	1.000

11.4.3 Problem 3

The exclusive or is a binary logical operation, described in Table 11.3. In this third example, the weights of a benchmark XOR neural network, represented in Fig. 11.8, are approximated using both the DSO and PSO algorithms.

This network has three layers:$L_1 = \{x_1, x_2\}, L_2 = \{z_1, z_2\}$ and $L_3 = \{y\}$. For every node of the the hidden and output layer, we have:

$$z_j = f\left(\sum_{i=1}^{2} w_{ij}^1 x_i + w_j^2\right), j = 1, 2,$$

$$y = f\left(\sum_{i=1}^{2} w_{i1}^2 x_i + w_1^3\right),$$

Table 11.3 XOR operation

x_1	x_2	XOR $\rightarrow y^-$
0	0	0
0	1	1
1	0	1
1	1	0

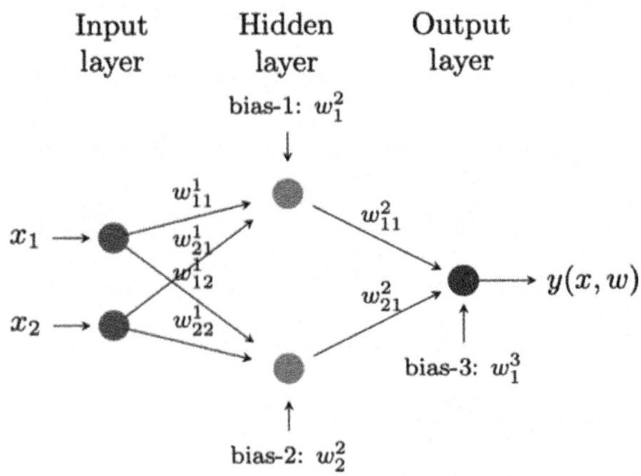

Fig. 11.8 Neural network for the xor problem

Table 11.4 Example 3: Comparison between the two algorithms

	Solution: P_i	Performance
DSO	(−0.807, −0.529)	0.999623
PSO	(7.969, 4.560)	0.606531

where f is the activation function and with $\sum_{i=1}^{2} w_{ij}^{1}x_i + w_j^2, j = 1, 2$ and $\sum_{i=1}^{2} w_{i1}^{2}x_i + w_1^3$ as the separable planes. w_{ij}^{L} represents the weight from node−i in layer−L to node−j in layer−(L + 1) and w_i^L is the bias of node−i in layer−L. Also

$$y(x, w) = w_{11}^{2}z_1 + w_{21}^{2}z_2. \tag{11.15}$$

The DSO algorithm is applied to find the optimum weights of the neural network. To do this, 100 $P_i = w^i \in \mathbb{R}^9$ are randomly generated with $w^i = \left(w_1^2, w_2^2, w_1^3, w_{11}^1, w_{12}^1, w_{21}^1, w_{22}^1, w_{11}^2, w_{21}^2\right)^i$. The discrepancy between the output calculated with the guessed weights using formula (11.15) and $\overline{y}$ is calculated for every P_i:

$$E = (\overline{y} - y) \cdot (\overline{y} - y)^T \tag{11.16}$$

where T is the transpose operation. The performance of every particle is then found using D(i) = e − E and we select maxi D(i) along the iterations. Iteratively, the weights are rectified until either E is zero or the maximum of the performance is obtained.

When search algorithms are applied to this type of problems, the procedure is very likely to get trapped into local maxima. The dark-matter particles seem to have a relevant role in this optimisation problem since they help to explore more thoroughly the search space, but they also seem to entrap solutions using their repulsive forces. The network was able to learn with the XOR operation result is very acceptable as it can be seen in Table 11.4, where a solution obtained by the DSO is frankly better than the one of the PSO. This illustrates the potential of the DSO algorithm.

The algorithms have been implemented in Matlab by MathWorks R2021 and run in a MacBook Air with processor 1.1 GHz Quad-Core Intel Core i5 and Ram capacity 8 GB and frequency 3733 MHz.

11.5 Conclusion and Future Work

In this work, a version of the GSA is presented where, besides gravitational particles, dark-matter particles are also included. This feature is useful in overcoming some weak points of the standard version of the algorithm, widely mentioned in the literature. Among these there is the danger of the algorithm getting trapped into a local optimum. This can be grave in a problem like function 1 but not so grave in a problem like function 2. The dark-matter particles do contribute for a more thorough search of the feasible space, making sure that the absolute global optimum is attained in the end of the procedure, as illustrated in Problem 2. Indeed, in the first example, the solution could get away from the crown of solutions and reach the peak. Moreover, the algorithm was able to understand the existence of an optimal solution as well as the set of sub-optimal ones. In the second example, it was able to find the absolute maximum even when they are so similar and devolve a solution very comparable to the one obtained by the PSO. In the third example, DSO was able to find the optima weights of the feedforward neural network and, this time, the solution found by the DSO is frankly better than the one of the PSO, demonstrating, therefore, the potential of the DSO as a learning algorithm. One can also see that the algorithm works better with many particles, however, in these simulations, we used only 100 particles to test its robustness. The use of more particles will contribute to a more thorough search of the space as well as the improvement of convergence.

In the future, we plan to perform a thorough appraisal of the algorithm, namely obtaining some statistical parameters, studying convergence rate, robustness and using diverse benchmark problems. In perspective is also the application of this algorithm to some real problems.

Acknowledgements Work financed by FCT—Fundação para a Ciência e a Tecnologia under project: (i) UIDB/04033/2020 for the first author. (ii) UIDB/00048/2020 for the second author.

References

1. Griva, S.I., Sofer, A.: Linear and Nonlinear Optimisation, 2nd edn. Siam, Philadelphia, PA (2009)
2. Marti, K.: Stochastic Optimisation Methods. Springer, Heidelberg, Germany (2005)
3. Michalewicz, Z.: Genetic Algorithms + Data Structures = Evolution Programs. Springer, Berlin (1996)
4. Zhigljavsky, A., Zilinska, A.: Stochastic Glocal Optimisations. Springer Optimisation and its Applications, New York (2008)
5. Conn, K.A.R., Vicente, L.N.: Introduction to Derivative-Free Optimization. MPS-SIAM Book Series on Optimization. Siam, Philadelphia, PA (2009)
6. Rhinehart, R.R.: Engineering Optimization: Applications, Methods, and Analysis. ASME Press, eBook (2018)
7. Cassioli, A.: A tutorial on black–box optimization. https://www.lix.polytechnique.fr/dambrosio/blackboxmaterial/Cassioli1.pdf (2013)
8. García, J.M., et al.: Genetic algorithms for mathematical optimization. J. Phys.: Conf. Ser. **1448**, 012020 (2020)
9. Santoso, G.K.V.S., Saraf, N.: Intelligent techniques for planning distributed generation systems. In: IEEE Power Engineering Society General Meeting (2007)
10. Glover, F.: Tabu search—part i. ORSA J. Comput. **1**(3), 190–206 (1989)
11. Glover, F.: Tabu search—part ii. ORSA J. Comput. **2**(1), 4–32 (1990)
12. Dorigo, M., Stutzle, T.: Ant Colony Optimization. The MIT Press, Cambridge (2004)
13. Geem, J.H.K.Z.W., Loganathan, G.: A new heuristic optimization algorithm: harmony search. Simulation **76**(2), 60–68 (2001)

14. Rashedi, H.N.E., Saryazdi, S.: A gravitational search algorithm. Inf. Sci. **179**, 2232–2248 (2009)
15. Ab Wahab, S.M.N., Atyabi, A.: A comprehensive review of swarm optimization algorithms. PLoS ONE **10**(5), e0122827 (2015)
16. Kennedy, J., Eberhart, R.: Particle swarm optimization (PSO). In: Proceedings of the IEEE International Conference on Neural Networks, Perth, UK, no. 4, pp. 1942–1948 (1995)
17. Kennedy, J.: Particle Swarm Optimization, pp. 760–766. Springer US, Boston, MA (2010)
18. Abdmouleh, A.Z., Ben-Brahim, L.: Review of optimization techniques applied for the integration of distributed generation from renewable energy sources. Renew. Energy **113**(3), 266–280 (2017)
19. Rashedi, H.N.E., Saryazdi, S.: Binary Gravitational Search Algorithm, vol. 9, pp. 727–745. Springer Science+ Business Media B.V. (2010). https://doi.org/10.1007/s11047-009-9175-3
20. Hassanzadeh, H.R., Rouhani, M.: A multi-objective gravitational search algorithm. In: 2nd International Conference on Computational Intelligence, Communication Systems and Networks, pp. 7–12 (2010). https://doi.org/10.1109/CICSyN.2010.32
21. Hu, X.H., Bai, Y.: Two kinds of classifications based on improved gravitational search algorithm and particle swarm optimization algorithm. Adv. Math. Phys., 2131862 (2017). https://doi.org/10.1155/2017/2131862
22. Li, J., Dong, N.: Gravitational search algorithm with a new technique. In: 13th International Conference on Computational Intelligence and Security (CIS), pp. 516–519 (2017). https://doi.org/10.1109/CIS.2017.00120
23. Mittal, A.C.P.H., Tripathi, A., Pal, R.: Gravitational search algorithm: a comprehensive analysis of recent variants. Multimed. Tools Appl. **80**, 7581–7608 (2021). https://doi.org/10.1007/s11042-020-09831-4
24. Sabri, M.P.N.M., Mahmood, M.R.: An overview of gravitational search algorithm utilization in optimization problems. In: 2013 IEEE 3rd International Conference on System Engineering and Technology, Shah Alam, Malaysia, pp. 19–20, August 2013. https://doi.org/10.1109/ICSEngT.2013.6650144
25. Sabri, M.P.N.M., Mahmood, M.R.: An overview of gravitational search algorithm utilization in optimization problems. Int. J. Adv. Soft Comput. Appl. **5**(3), 1–39 (2013)
26. Rathee, N., Chhillar, R.S.: Gravitational search algorithm: a novel approach for structural test path optimization. J. Interdiscip. Math. **23**(2), 471–480 (2020). https://doi.org/10.1080/09720502.2020.1731960
27. Zheng, H.L.C., Wang, L.: An improvement of gravitational search algorithm. In: Yingmin, J.D.J., Zhang, W. (eds.) Proceedings of 2019 Chinese Intelligent Systems Conference, Singapore, pp. 490–503. Springer Singapore (2020)
28. Eldos, T., Al Qasim, R.: On the performance of the gravitational search algorithm. Int. J. Adv. Comput. Sci. Appl. **4**(8), (2013)
29. Khalaf, M., Al-Neamy, J.: Hybrid gravitational search algorithm and genetic algorithms for automated segmentation of brain tumours using feature based symmetric analysis. (IJC-SIS) Int. J. Comput. Sci. Inf. Secur. **11**(5), 30–38 (2013)
30. Duman, Y.S.S., Güvenc, U., Yörükeren, N.: Optimal power flow using gravitational search algorithm. Energy Convers. Manag. **59**, 86–95 (2012)
31. Geleta, D.K., Manshahia, M.S.: Gravitational search algorithm-based optimization of hybrid wind and solar renewable energy system. Comput. Intell. **Special Issue**, 1–27 (2020). https://doi.org/10.1111/coin.12336
32. Hosseinabadi, M.A.R., Rostami, A.S.: A new search algorithm for solving symmetric traveling salesman problem based on gravity. World Appl. Sci. J. **16**(10), 1387–1392 (2012)
33. Gu, X.W., Zhu, L.: A gravitational search algorithm for flow shop scheduling. CAAI Trans. Intell. Syst. **5**(5), 411–418 (2010)
34. Papa, S.J.P., Pagnin, A., Spadotto, A.: Feature selection through gravitational search algorithm. In: IEEE International Conference on Acoustics, Speech and Signal Processing (ICASSP), pp. 2052–2055 (2011)

35. Zhaoa, W.: Adaptive image enhancement based on gravitational search algorithm. Procedia Eng. **15**, 3288–3292 (2011)
36. Xiao, J., Cheng, Z.: Theories and applications dna sequences optimization based on gravitational search algorithm for reliable dna computing. In: Sixth International Conference on Bio-Inspired Computing. IEEE Computer Society (2011)
37. Rashedi, H.N.E., Saryazdi, S.: Filter modeling using gravitational search algorithm. engineering. Appl. Artif. Intell. **24**, 117–122 (2011)
38. de Almeida, B.S.G., Leite, V.C.: Particle swarm optimization: a powerful technique for solving engineering problems. In: Javier Del Ser, E.V., Osaba, E. (eds.) Swarm Intelligence - Recent Advances, New Perspectives and Applications. IntechOpen (2019)

Chapter 12
A Statistical Approach to Analyzing Turkish Morphology

Duygu Altinok

Abstract Morphological analyzers are essential components of Turkish language processing pipelines due to complex and rich morphology of Turkish language. Previous work usually focuses on two-level description of Turkish morphology, thus leads to lexicon and two-level rules oriented FST implementations. However, two-level based implementations are not robust to spelling errors and new loan words entering the Turkish lexicon. In this paper, we introduce a statistical approach to analyzing Turkish word forms by training and comparing two seq2seq models on our annotated dataset. We approach analyzing Turkish word forms as a machine translation problem, where the source sequence is a Turkish word form and the target sequence is a sequence of morphemes. Evaluating on three testsets of informal written language word forms, we show that our approach offers a robust approach to analyzing the Turkish words and proposes a strong baseline for sequential modeling of Turkish morphology.

Keywords Attention · seq2seq · Statistical morphological analysis · Turkish morphology

12.1 Introduction

Turkish morphology is agglutinative with rich inflectional and derivational morphotactics, an immense set of affixes and morphophonemic processes. Turkish morphology has been well-studied by [1–3]. Turkish word structures are formed by inflectional and derivational suffixes to root words. An example of word formation process is:

```
gitmeyecekmişçesine
```

which can be decomposed into root and suffixes as follows:

D. Altinok (✉)
Deepgram San Francisco, San Francisco, CA 94104, US
e-mail: duygu.altinok@deepgram.com

M. A. Laribi et al. (eds.), *Advances in Automation, Mechanical and Design Engineering*, Mechanisms and Machine Science 121,
https://doi.org/10.1007/978-3-031-09909-0_12

```
git+me+yecek+miş+çesine
```

where +'s indicate morpheme boundaries. This example translates to English as "As if (s)he is not going to leave," which is not only one word but a clause in English. Agglutinative morphology has the ability to encode in a single word the equivalent of whole clauses and sentences in languages with simpler morphology. Thus full morphological analysis is required before downstream Turkish language processing tasks. In the respect morphological analyzers are essential components of statistical and rule-based Turkish language processing pipelines.

Previous approaches for analyzing Turkish morphology usually focused on two-level morphology [4]. The first two-level description of Turkish morphology is presented by [5] and later adapted by [6] to train a stochastic finite-state transducer (SFST) on a 200 million word corpus. References [7, 8] put emphasis on availability and propose open-source alternatives morphological analyzers. Ozturel et al. [9] presents a Turkish morphology model that improves modeling morphotactics for lexical categories and offers a novel approach to represent named entities.

In two-level approach, besides the two-level rules large lexicons of words categorized by lexical categories are used. Then the model is implemented as an FST from the lexicon and the two-level rules. Consequently, NLP applications need to store a potentially large binary file. Also these models are not robust to misspelled words and loan translation words that come to Turkish from English words such as "tweet," "tivitlemek" (to tweet, verb) and "googlelamak" (to google, verb).

Dealing with those issues, we propose a statistical approach to Turkish morphological analysis. In this work we viewed morphological analysis as a machine translation problem, where the source language is Turkish word forms and the target language is the language of inflectional and derivational suffix tags and the root words with their part-of-speech (POS) tags. More specifically, we map a word to a sequence of words the form `(root, root POS tag, tag1, tag2, tag3...)`. Sequence-to-sequence modeling (seq2seq) has been successfully applied to various alignment problems including neural machine translation [10, 11], named entity recognition [12, 13] and speech recognition [14–16]. First, we implemented CharEncAtt, a character level encoder-decoder model with self-attention. The model architecture consists of a bidirectional Long Short-Term Memory (BiLSTM) layer [3] to encode the input words and a second BiLSTM to decode the output sequence. A self-attention layer is placed on top of the encoder BiLSTM to relate the input word form's characters. We evaluated our model with the overall accuracy by viewing the problem as a multi-label classification task. Our second model CharTransAtt is again a character level encoder-decoder model, but with Bahdanau attention [17] to relate the input sequence and the output sequence. We implemented the architecture proposed [18], an effective and popular seq2seq architecture for neural machine translation. The model includes a BiLSTM encoder that encodes the input word forms and a BiLSTM decoder which generates one target sequence element at one step. The target sequence is a sequence of characters and sequence tags, the word lemma is

Table 12.1 Example entries from the lexicon

Word form analysis	
ev	ev,N,num.sg,case.nom
eve	ev,N,num.sg,case.dat
evimde	ev,N,num.sg,p1s,case.loc
okulumda	okul,N,num.sg,p1s,case.loc
isme	isim,N,num.sg,case.dat
koyun	koyun,N,num.sg,case.nom
koyun	koy,V,imper.2p.pl.short

generated character by character. The global attention layer lies between the encoder BiLSTM and the decoder LSTM [19].

One of the challenges imposed by previous approaches is needing to store large binary files during the runtime, we keep the statistical model space-efficient by keeping the embedding layer dimensions small by fixing the first dimension as the number of characters in Turkish alphabet. Also our statistical approach overcomes the challenge of recognizing loan words and misspelled words.

Evaluating on datasets of native Turkish words, misspelled words and loan translation words, we show that our model offers a robust approach to analyzing the Turkish word forms and a strong baseline for sequential modeling of Turkish morphology.

12.2 Turkish Morphological Dictionary

Our aligned dictionary consists of word forms and their morphological analysis. An entry is a 4-tuple of word form, its root, root's POS tag and a sequence of the suffix tags (see Table 12.1).

A morphological analysis string represents the lexical form of the word form [5]. The preceding word form "evimde" can be decomposed as "ev+im+de" for an example. "num.sg" tag indicates a singular number suffix which corresponds to empty string, "pls" is first person possessive suffix that corresponds to the surface realization "im" and "case.loc" is locative case suffix which realized the surface form "de." Each suffix tag corresponds to a suffix from Turkish grammar, a suffix can be possibly correspond to empty string. A comprehensive list of suffixes can be found in [20].

We followed a semi-automatic process to build our dictionary. First we collected word forms from the Turkish Wikimedia dump[1] and these word forms are annotated by a human annotator. Then we extracted a list of root words and a list of suffixes from the annotated word forms. We generated new word forms by making combinations of the root words and the suffixes that we obtained from the previous step. Finally we make a pass over the new word-suffix combinations with a character level seq2seq

[1] https://www.dumps.wikimedia.org/trwiki/

model for modeling the morphological processes to build the correct word form. Figure 12.1 explains the generation of new word forms.

Our dictionary contains 403,634,549 entries. Word forms are grouped into text files by their root POS tag. The distribution of word forms by root POS tag is shown in Table 12.2.

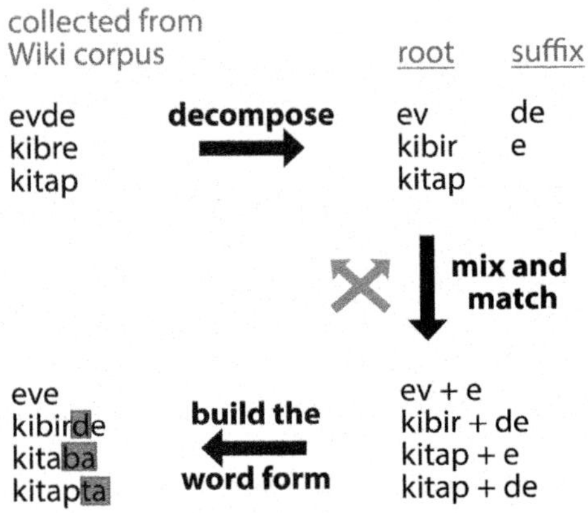

Fig. 12.1 Example generation of new word forms from Wikimedia corpus word forms. We collect three words forms: "evde," "kibre" and "kitap." Annotators decompose them into root and suffixes as "ev+de," "kibir+e" and "kitap." Root word "kibir" loses its second vowel during admitting the dative suffix "-e." We cross product the list of roots ["ev," "kibir," "kitap"] and a list of suffixes ["de," "e"] we collected from the previous step. We obtain new lexical forms "ev+e," "kibir+de," "kitap+e" and "kitap+ta." In order to build the correct word forms from the lexical forms, we make a pass over the lexical forms with a characater level seq2seq model

Table 12.2 Distribution of word forms by the root POS tags. We used Universal Dependencies (UD) POS tagset [21] with minor enhancements. For the complete set of POS tags and suffix tags we use, refer to the link https://www.github.com/turkish-nlp-suite/turkish-morph-lexicon/blob/main/tagset.md

POS	Description	Number
ADJ	Adjective	224,719,573
ADV	Adverb	1,661
CONJ	Conjunction	69
DET	Determiner	211,189
POSTP	Postposition	1,914
PRON	Pronoun	326,735
N	Noun	19,861,063
N.proper	Proper noun	88,395,306
N.geo	Geographical name	5,353,438
V	Verb	64,753,483
INTJ	Interjection	118

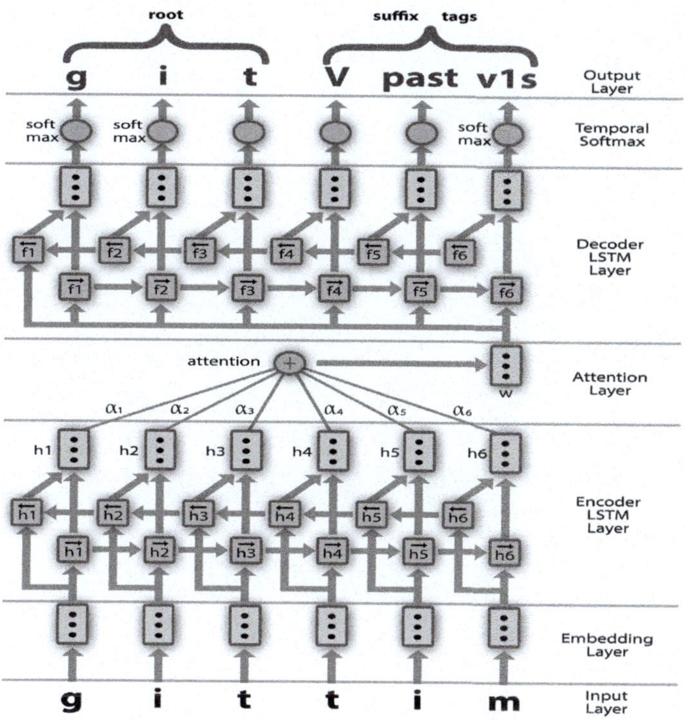

Fig. 12.2 Character level encoder-decoder model with self-attention

We also included the list of suffix tags used and an annotation guideline with example word forms and analyses. Turkish Morphology Dictionary is open source and freely available.[2]

12.3 Models

Both our models take a word form $x = c_1, \ldots, c_N$, where each c_i is a Turkish alphabet character and outputs corresponding sequence$y = y_1, \ldots, y_M$, where each y_i is either a character or a suffix tag. The models output word lemma character by character, followed by the POS tag of the word lemma and a sequence of suffix tags. An example input word form and the corresponding output sequence is shown in Figs. 12.2 and 12.3.

---

[2] https://www.github.com/turkish-nlp-suite/turkish-morph-lexicon.

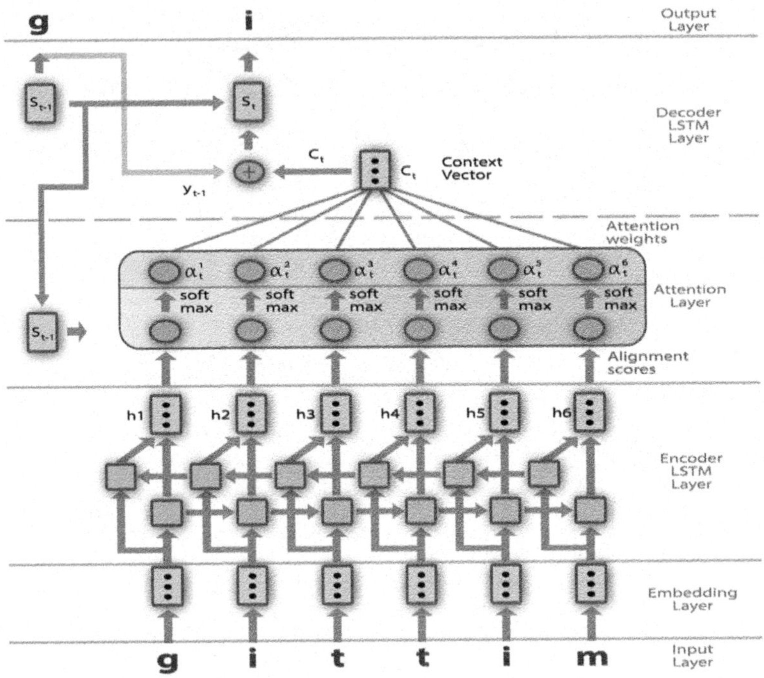

Fig. 12.3 Character level encoder-decoder model with global attention

12.3.1 *CharEncAtt*

In this section we propose our CharEncAtt model in detail. As shown in Fig. 12.2, the model contains six components:

1. Input layer: inputs word form to this model
2. Embedding layer: maps each character to a vector representation
3. Encoder LSTM layer: encodes character sequences with a BiLSTM
4. Attention layer: calculates attention over the hidden states of the BiLSTM in step (1) and produce a word form vector
5. Decoder LSTM layer: decodes the word form vector with a BiLSTM
6. Output layer: two time distributed dense layers to generate a root word and a sequence of morphemes.

Details of these components will be presented in this section.

Word Representation. In order to leverage character level information, we use character embeddings to represent the input word forms. Let C denote the set of characters of the Turkish alphabet, together with digits and hyphen character respectively. Let $\oplus$ denote the vector concatenation operator as well. The model parameters associated with the character representations are:

$$\textbf{Charembedding}: e_c \in R^{100} \text{ for each } c \in C$$

We encode a character vector sequence $(c_1, \ldots, c_N)$ with a BiLSTM in order to capture information from both left and right sides of the sequence. The hidden state of the BiLSTM can be stated as follows

$$\overrightarrow{h}_i = \overrightarrow{LTSM}\left(\overrightarrow{h}_{i-1}, e_{c_i}\right) \tag{12.1}$$

$$\overrightarrow{h}_i = \overleftarrow{LTSM}\left(\overleftarrow{h}_{i+1}, e_{c_i}\right) \tag{12.2}$$

$$h_i = \overrightarrow{h}_i \oplus \overleftarrow{h}_i \tag{12.3}$$

where $\overrightarrow{h}$ and $\overleftarrow{h}$ are the hidden states of the forward LSTM and the backward LSTM respectively. Hidden dimension of the BiLSTM is also 100, hence each character c_i is represented by a 200 dimensional vector h_i, which is concatenation of the forward and backward hidden states.

Self-Attention. Attentive seq2seq models have demonstrated huge success in a broad range of tasks including question answering, POS tagging, semantic role labeling, speech recognition and machine translation. We applied self-attention to learn the dependencies between any two input characters and encapsulate the morpheme level information such as morpheme boundaries, morphotactics and morphophonemic processes. We implemented a feed-forward attention as described in [22]. Specifically, the self-attention layer takes the hidden states of encoder layer h_i through an activation layer to get a hidden representation u_i. Then, importance of a character is measured by calculating the similarity between u_i and the character level context vector u_c. Then we normalize the similarity score through a softmax function to get importance weight α_i. Finally we compute the word form vector w as a weighted sum of the encoder hidden states with attention weights. The context vector u_c is a trained parameter vector and can be viewed as a high-level representation of importance over the characters.

$$u_i = tanh(h_i) \tag{12.4}$$

$$\alpha_i = \frac{\exp\left(u_i^\mathsf{T} u_c\right)}{\sum_k \exp\left(u_k^\mathsf{T} u_c\right)} \tag{12.5}$$

$$w = \sum_i \alpha_i h_i \tag{12.6}$$

The self-attention layer lies on top of the encoder LSTM layer, refer to Fig. 12.2 for the details.

Morpheme Sequence Generation. To estimate the morpheme sequence y corresponding to an input word form, we again use a BiLSTM for training a decoder, where the encoded word form knowledge w is fed into the input of the model. Then we apply a softmax layer to output of the BiLSTM at each temporal time step to classify the root word's characters and the suffix tags:

$$\overrightarrow{f}_j = \overrightarrow{LTSM}\left(\overrightarrow{f}_{j-1}, w\right) \tag{12.7}$$

$$\overleftarrow{f}_j = \overleftarrow{LTSM}\left(\overleftarrow{f}_{j+1}, w\right) \tag{12.8}$$

$$f_j = \overrightarrow{f}_j \oplus \overleftarrow{f}_j \tag{12.9}$$

$$y_j = softmax\left(f_j\right)\forall j \in \{1, \ldots, M\} \tag{12.10}$$

where M is the output sequence length.

12.3.2 *CharTransAtt*

In this section, we describe our second model, CharTransAtt in detail. The model contains one more layer than CharEncAtt, an additional input layer to the decoder LSTM:

1. Encoder Input layer: inputs word form to the encoder layer.
2. Decoder Input layer: inputs a target sequence element to the decoder layer at each timestep during training. During inference, this layer inputs the output of the previous time step to the decoder.
3. Embedding layer: maps each character and sequence tag to a vector representation.
4. Encoder LSTM layer: encodes character sequences with a BiLSTM.
5. Global-Attention layer: uses the hidden states of the BiLSTM in step (4) and decoder LSTM's current hidden state to calculate attention scores and attention weights to produce a context vector.
6. Decoder LSTM layer: decodes the current timestep's output with a BiLSTM. Decoder LSTM computes the decoder next hidden state and decoder output from the decoder input in step (2), context vector in step (5) and its previous hidden state.
7. Output layer: softmaxes the decoder LSTM's output to generate the current timestep's final output.

CharTransAtt decodes one output element at each timestep, where CharEncAtt outputs the result sequence at once.

Encoder and Decoder Input Layers. The encoder input layer inputs word forms to the encoder layer. The decoder input layer is different from the encoder input layer in the sense that, at each time step it inputs one target sequence element (a character or a suffix tag) to the decoder layer.

Embedding Layer. We use the same embedding layer for both encoder inputs and decoder inputs to enable encoder and decoder LSTMs to share the same representation for the characters. The suffix tag embeddings has no contribution to the encoder LSTM weights, as the encoder processes word forms, hence only alphabetic characters. We again use 100 dimensional embeddings to represent the input characters and the suffix tags:

$$\textbf{Enrichedembedding} : e_d \in R^{100} \text{ for each } d \in D$$

where D is the set of suffix tags combined with set of characters of the Turkish alphabet together with digits and hyphen character.

Encoder. The Encoder layer encodes a character vector sequence $c_1, \ldots, c_N$ with a bidirectional LSTM. The hidden state of the Encoder BiLSTM can be formulated as follows:

$$\overrightarrow{s}_t = \overrightarrow{LTSM}(\overrightarrow{s}_{t-1}, e_{c_i}) \tag{12.11}$$

$$\overleftarrow{s}_t = \overleftarrow{LTSM}(\overleftarrow{s}_{t+1}, e_{c_i}) \tag{12.12}$$

$$s_t = \overrightarrow{s}_t \oplus \overleftarrow{s}_t \tag{12.13}$$

We used 256 as the hidden dimension, hence forward and backward LSTMs map each character vector to a 256 dimensional vector. Therefore each character c_i is represented by a 512 dimensional vector s_t which is concatenation of the forward and the backward hidden states.

Global-Attention. Global and local attention mechanisms were introduced in [18], where global attention attends to all elements of the input sequence and local attention considers only a subset of the input sequence. We applied global attention to learn the dependencies between input characters and output sequence elements-characters and suffix tags, to encapsulate the morphological information. We implemented Bahdanau attention as described in [17]. At each decoder step t, we feed all the encoder LSTM hidden states $h_{k=1}^{N}$ and the decoder LSTM's previous hidden state s_{t-1} (just before emitting y_t) to the global attention layer. For each encoder state h_k, we compute its relevance score for the decoder state s_{t-1}. Then we compute the attention weights by applying a softmax function to attention scores. Finally the context vector c is computed as the weighted sum of the encoder states with attention weights. The details of the computation is as follows:

$$score(s_t, h_k) = s_{t-1}^{\mathsf{T}} W h_k \ \ for\ \mathbf{k} = 1, \ldots, \boldsymbol{N} \tag{12.14}$$

$$\alpha_t^{(k)} = \frac{exp(score(s_{t-1}, h_k))}{\sum_{i=1}^{N} exp(score(s_{t-1}, h_k))} \quad for\ \mathbf{k} = 1, \ldots, N \tag{12.15}$$

$$c_t = \sum_{k=1}^{N} \alpha_t^k h_k \tag{12.16}$$

where W is a trainable weight matrix. The global attention layer lies between the encoder layer and decoder layer, refer to Fig. 12.3.

Decoder. The decoder LSTM is a bidirectional LSTM that produces an output vector and a hidden state vector at each timestep. The current hidden state vector s_t is calculated from the decoder layer's previous output y_{t-1}, the current context vector c_t and the previous hidden state s_{t-1}:

$$s_t = f(s_{t-1}, y_{t-1}, c_t) \tag{12.17}$$

First we concatenate the context vector c_t and the previous output element's embedding vector $e_{y_{t-1}}$. Then we feed this combined vector and the previous hidden state s_{t-1} to the decoder BiLSTM to get the current hidden state s_t and the current output vector o_t. We apply a softmax function to the current output vector o_t to obtain y_t, the final output of the current timestep, which corresponds to a vocabulary item—an alphabet character or a suffix tag.

$$s_t, o_t = BiLSTM\left(s_{t-1}, c_t \oplus e_{y_{t-1}}\right) \tag{12.18}$$

$$y_t = softmax(o_t) \tag{12.19}$$

Global attention layer lies on top of the encoder LSTM layer; please refer to Fig. 12.3 for the architecture overview.

12.4 Experiments

In this section, we describe settings of experiment parameters in detail. We implemented both model architectures in Keras framework because of its flexibility and the community support.[3]

[3] Implementations can be found at https://www.github.com/turkish-nlp-suite/turkish-morph-lexicon/tree/main/keras_code

Table 12.3 The number of word forms in the three datasets

	Lexicon	Misspelled	Loan
Train	8 M	17 K	103 K
Dev	1 M	1 K	15 K
Test	1 M	2 K	30 K

12.4.1 Datasets

We evaluated our models on three test sets: (1) words from the Turkish lexicon (2) misspelled lexicon words (3) loan translation words. Lexicon words dataset is subsampled from Turkish Morphological Dictionary we described in Sect. 12.2 to reduce the training time. Loan translation words test set is compiled by a human annotator, including popular loan translation words such as "stalk" (N), "emoticon" (N), "tweetlemek" (V, to tweet), "stalklamak" (V, to stalk), "likelamak" (V, to click the like button). We generated the misspelled words from the lexicon words by making random insertions, deletions and substitutions. Table 12.3 shows the statistics of the datasets.

There are 126 suffix tags in total, 41 of them being copulas. The full list of the tag set can be viewed at the link below.[4]

12.4.2 Implementation Settings

For evaluation of CharEncAtt, we used categorical accuracy as evaluation metric in our experiments. For training CharEncAtt model, we use mini-batch stochastic gradient descent with batch size 64. We use the ADAM optimizer [23] with default parameters (learning rate 0.001, $\beta_1 = 0.9$, $\beta_2 = 0.999$) for the optimization. The model is trained 20 epochs on each batch. Encoder BiLSTM's hidden layer size is 100, where decoder BiLSTM's hidden layer size is 300. The dropout rate is set to 0.3 to prevent overfitting.

For trainable parameters initialization, we use Xavier initializer [24] to initialize parameters, including the character embeddings used in our experiment. The model size is 76 M.

Similarly, for the evaluation of CharTransAtt, we used categorical accuracy as the evaluation metric. The hidden layer size of both the encoder and decoder LSTMs are 256. We didn't use any sequential dropout in the LSTMs. We used the ADAM optimizer with default parameters and set the batch size as 64. We trained the model 20 epochs on each batch. For initializing the trainable parameter, we again used Xavier initializer.

[4] https://www.github.com/turkish-nlp-suite/turkish-morph-lexicon/blob/main/tagset.md.

Table 12.4 The model performance on the test sets. Performance measure is categorical accuracy

	Lexicon	Misspelled	Loan
CharEncAtt	0.99	0.96	0.89
CharTransAtt	0.98	0.95	0.91

12.4.3 Results and Discussion

In this section, we present the results and compare the results of the architectures CharEncAtt and CharTransAtt. We also exhibit the visualizations of attention weights for each model to gain more insights.

Results. Table 12.4 shows the classification accuracy of each model on the three test sets. Our models performed best on the Turkish lexicon words as expected. Both model performances slightly decline on the misspelled words. In some cases of the loan translate words are easier to segment into morphemes. For an example the word "googlelamak" gives a hint to the model by the suffix "-la" (a derivational suffix that converts nouns to verbs), followed by the suffix "-mak" (infinitive form suffix that converts verbs to verbs, literal translation is "to"). The root word's ("google") last vowel is "e," followed by the suffix "la" which violates the vowel harmony rules [1]. Collecting these hints, the model can infer where the root boundary occurs and "googlelamak" should be a verb derived from the noun "google." Not all the translated noun words straightforward to infer, for an example the word "stalk" does not give any immediate indications of its syntactic category or the root word. The word "stalktan" ("stalk" noun in ablative case) can be inferred as a noun because it admitted the ablative case suffix, where the word "stalk" does not provide any information about its syntax category or the root boundary at all.

Figure 12.4 shows the loss and accuracy per epoch graphs for our first model, CharEncAtt.

The preceding figure exhibits that the training and test performances are similar, hence we deduce that no overfitting happened during training. We compare these graphs with the loss and accuracy per epoch graphs of CharTransAtt, given by Fig. 12.5.

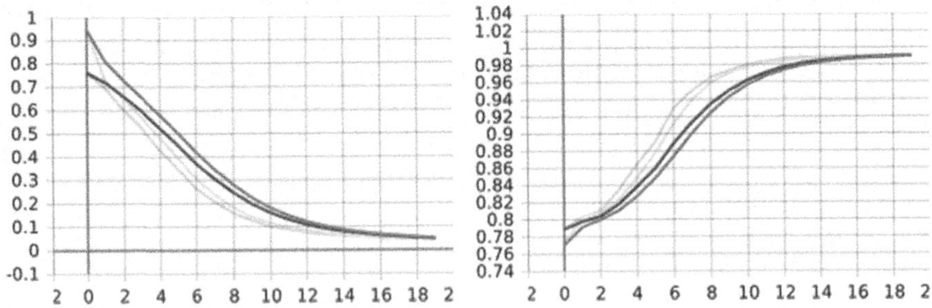

Fig. 12.4 Performance of CharEncAtt model on the training and test sets. The blue line shows the test set loss/accuracy and the orange line shows training set loss/accuracy

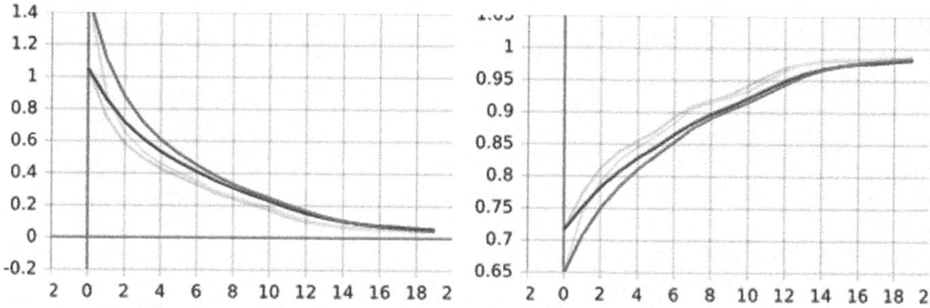

Fig. 12.5 Performance of CharTransAtt model on the training and test sets. The blue line shows the test set loss/accuracy and the orange line shows training set loss/accuracy

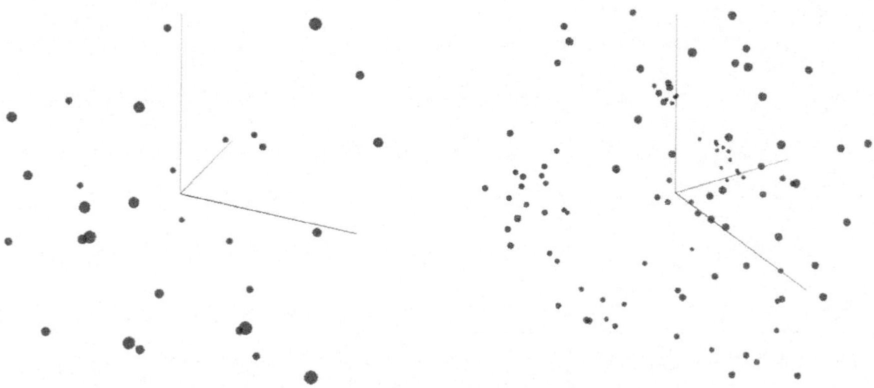

Fig. 12.6 CharEncAtt and CharTransAtt embeddings side by side

The character embeddings of CharEncAtt and CharTransAtt's enriched character and suffix tag embeddings are shown in Fig. 12.6. CharEncAtt has a vocabulary of size 44, which corresponds to size of the Turkish alphabet and special characters. CharTransAtt has a bigger vocabulary size, as input sequences and target sequences share the same embedding layer. The total vocabulary size is 170, summation of the Turkish alphabet's size and set of suffix tags' size.

CharEncAtt and CharTransAtt performed very similarly on all three datasets. CharEncAtt slightly outperformed CharTransAtt on Turkish lexicon words and misspelled words. However, CharTransAtt performed better on loan translate words. This might originate from the capability of the global attention to capture the input word form and output morphological analysis, global attention layer can attend the morpheme boundaries more carefully.

Visualization of Attention. In this section we provide the visualizations of attention weights for each model. Figure 12.7 shows the self-attention weights of each

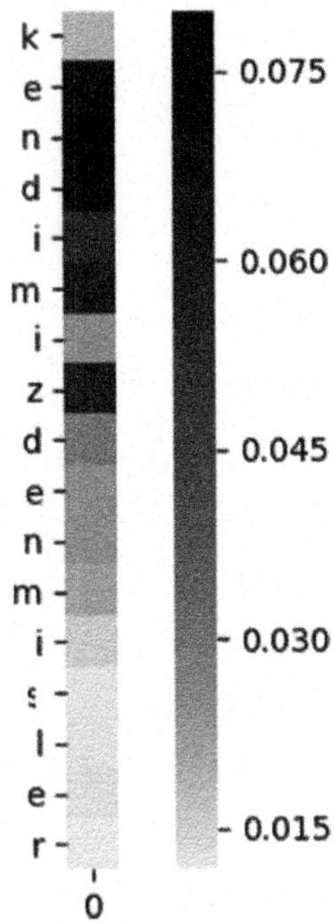

Fig. 12.7 Importance of each character of the input word form with respect to the context representation. Darker colors indicate more importance. This word can be broken down into its morphemes as "kendi+miz+den+miş+ler." The character "m" has a significant weight due to the fact that it lies on the morpheme boundary. Same applies to the character "z," it is the last character of the same morpheme

character for the example word form "kendimizdenmişler," generated by CharEncAtt. Each input character's importance is measured by calculating the similarity between this character and a trainable, character level context vector.

Figure 12.8 shows CharTransAtt's input sequence-output sequence attention weights for the same example word form "kendimizdenmişler" and its morphological analysis. Each output sequence element attends to input sequence characters. From the figure, we see that the consonants of the word lemma "k," "n," and "d" attends only to themselves in the input sequence. The situation is different for the lemma vowels, the characters "e" and "i" attend to all the input vowels "e," "i," "i," "e," "i" and "e," following the vowel harmony rule of Turkish phonology [20]. Looking at the second row of Fig. 12.8, "e" attends to itself with the most attention weight, then it attends to the next vowel to right "i." The attention fades as the distance to "e" increases. Suffix tags follows a different pattern. The lemma's lexical category "PRON.reflex.3p.sg" tag attends all of the lemma characters "k," "e," "n," "d" and "i" in the input word form. The suffix tag "p1p" attends all the characters of the corresponding suffix "m," "i" and "z." Also, the characters "i" and "d"- left and right to the suffix "miz"—are also attended with a less amount of attention. Reason for this might be the Turkish morphophonemic rules, which always occur at the suffix boundaries [20]. Rest of the suffix tags behaves similarly and create attention blocks around the corresponding suffix characters.

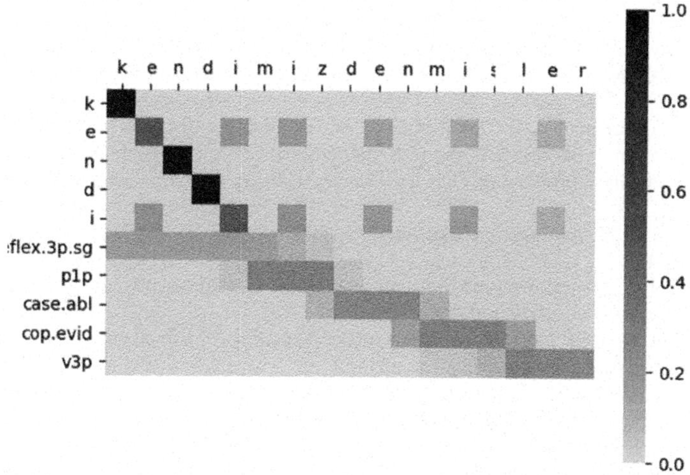

Fig. 12.8 Attention between the example word form "kendimizdenmişler" and its morphological analysis. Darker colors indicates more attention. Each morphological tag attends to input word form's characters

12.5 Conclusion

In this paper, we brought a statistical approach to analyzing Turkish word forms. We proposed two seq2seq models, CharEncAtt—a character level encoder-decoder model with self-attention and CharTransAtt—a character level encoder-decoder model with global attention to analyze the Turkish word forms. Our models do not depend on large root word lexicons and robust to analyzing misspelled word forms and loan words. The effectiveness of both CharEncAtt and CharTransAtt are demonstrated by evaluating the model on three test sets. We visualized the attention weights of each model to exhibit how the models explored the rules of Turkish morphology with a purely statistical approach.

References

1. Oflazer, K., Gocmen, E., Bozsahin, C.: An Outline of Turkish Morphology (1994)
2. Oflazer, K., Saraçlar, M.: Turkish Natural Language Processing. Springer International Publishing (2018). https://www.books.google.de/books?id=D-5lDwAAQBAJ
3. Chiu, J.P.C., Nichols, E.: Named entity recognition with bidirectional LSTM-CNNs. Trans. Assoc. Comput. Linguist. **4**, 357–370 (2016). https://doi.org/10.1162/tacl_a_00104
4. Koskenniemi, K.: Two-Level Model for Morphological Analysis. In: IJCAI (1983), pp. 683–685
5. Oflazer, .: Two-level Description of Turkish Morphology. Utrecht, The Netherlands, (1993). https://www.aclweb.org/anthology/E93-1066
6. Sak, H., Güngör, T., Saraçlar, M.: A stochastic finite-state morphological parser for Turkish. In: Proceedings of the ACL-IJCNLP 2009 Conference Short Papers, USA, 2009, pp. 273–276

7. Çöltekin, C.: A freely available morphological analyzer for Turkish. Valletta, Malta (2010). https://www.lrec-conf.org/proceedings/lrec2010/pdf/109_Paper.pdf
8. Akın, A.A., Akın, M.D.: Zemberek, an open source NLP framework for Turkic Languages (2007)
9. Ozturel, A., Kayadelen, T., Demirsahin, I.: A syntactically expressive morphological analyzer for Turkish. In: Proceedings of the 14th International Conference on Finite-State Methods and Natural Language Processing, Dresden, Germany (2019), pp. 65–75. https://doi.org/10.18653/v1/W19-3110
10. Stahlberg, F.: Neural machine translation: a review. CoRR (2019). http://www.arxiv.org/abs/1912.02047
11. Tan, Z., et al.: Neural machine translation: a review of methods, resources, and tools. CoRR (2020). https://www.arxiv.org/abs/2012.15515
12. Li, J., Sun, A., Han, J., Li, C.: A survey on deep learning for named entity recognition. CoRR (2018). http://www.arxiv.org/abs/1812.09449
13. Sutskever, I., Vinyals, O., Le, Q.V.: Sequence to sequence learning with neural networks. CoRR (2014). http://www.arxiv.org/abs/1409.3215
14. Bahdanau, D., Chorowski, J., Serdyuk, D., Brakel, P., Bengio, Y.: End-to-end attention-based large vocabulary speech recognition. In: 2016 IEEE International Conference on Acoustics, Speech and Signal Processing (ICASSP), pp. 4945–4949 (2016). https://doi.org/10.1109/ICASSP.2016.7472618
15. Chiu, C.-C., et al.: State-of-the-art speech recognition with sequence-to-sequence models. CoRR (2017). http://www.arxiv.org/abs/1712.01769
16. Nassif, A.B., Shahin, I., Attili, I., Azzeh, M., Shaalan, K.: Speech recognition using deep neural networks: a systematic review. IEEE Access **7**, 19143–19165 (2019). https://doi.org/10.1109/ACCESS.2019.2896880
17. Bahdanau, D., Cho, K., Bengio, Y.: Neural machine translation by jointly learning to align and translate. CoRR (2015)
18. Luong, M.-T., Pham, H., Manning, C.D.: Effective approaches to attention-based neural machine translation. CoRR (2015). http://www.arxiv.org/abs/1508.04025
19. Hochreiter, S., Schmidhuber, J.: Long short-term memory. Neural Comput. **9**, 1735–1780 (1997). https://doi.org/10.1162/neco.1997.9.8.1735
20. Göksel, A., Kerslake, C.: Turkish: A Comprehensive Grammar. Routledge (2005). https://www.books.google.de/books?id=7fXCKZmee8QC
21. Petrov, S., Das, D., McDonald, R.: A universal part-of-speech tagset. In: Proceedings of the Eighth International Conference on Language Resources and Evaluation (LREC'12), Istanbul, Turkey, May 2012, pp. 2089–2096. https://www.lrec-conf.org/proceedings/lrec2012/pdf/274_Paper.pdf
22. Yang, Z., Yang, D., Dyer, C., He, X., Smola, A., Hovy, E.: Hierarchical attention networks for document classification. In: Proceedings of the 2016 Conference of the North American Chapter of the Association for Computational Linguistics: Human Language Technologies, San Diego, California, June 2016, pp. 1480–1489. https://doi.org/10.18653/v1/N16-1174
23. Kingma, D.P., Ba, J.: Adam: A Method for Stochastic Optimization (2017)
24. Glorot, X., Bengio, Y.: Understanding the difficulty of training deep feedforward neural networks. In: Proceedings of the Thirteenth International Conference on Artificial Intelligence and Statistics, Chia Laguna Resort, Sardinia, Italy, May 2010, vol. 9, pp. 249–256. http://www.proceedings.mlr.press/v9/glorot10a.html

Chapter 13
Achievements of Artificial Intelligence in the Past and During the COVID-19 Era to Tackle Deadly Diseases

Zineb Dafir and Said Slaoui

Abstract Processing big data based on machine learning tools and deep learning models have become crucial in finding solutions to various problems in different fields. In particular, artificial intelligence systems play a key role in improving performance and increasing the speed of a clinical decision for a given clinical issue. In that respect, the outbreak of the virus caused by SARS-CoV-2 across the world has caused panic and unrest among people which required the rapid intervention of bioinformatics researchers and health professionals to discover the relevant clinical treatment. The main concern of the paper is to emphasize the usefulness of artificial intelligence and Big Data analytic for treating diseases. Thus, it discusses the past success of artificial intelligence and big data analytic for healthcare applications. It also presents the current advancements in machine learning tools as well as deep learning models for processing data sets related to the COVID-19 virus. In addition to that, the proposed work puts emphasis on the requirements of artificial intelligence models to make suitable clinical decisions. And finally, it gives a proposed analysis of the main factors that help decrease the spread of a pandemic in the world.

Keywords Artificial intelligence · Machine learning algorithms · Big data analytic · SARS-CoV-2 · Covid-19

13.1 Introduction

In an unprecedented way, the whole world is witnessing disturbance and much drastic change in different sectors, due to the propagation of a deadly epidemic that started in China and then quickly spread around the world. This global crisis made most of the countries take crucial decisions to limit its spread including declaring a state of emergency, full lock-downs, closing airports, forcing travel restrictions, and completely closing their borders. Parallel to these decisions, the majority of countries have experienced radical changes in several sectors. Indeed, some sectors have been absolutely

Z. Dafir (✉) · S. Slaoui
Mohammed V University in Rabat, 4 Avenue Ibn Battouta, B.P. 1014 RP, Rabat, Morocco
e-mail: zineb.dafir@um5s.net.ma

M. A. Laribi et al. (eds.), *Advances in Automation, Mechanical and Design Engineering*, Mechanisms and Machine Science 121,
https://doi.org/10.1007/978-3-031-09909-0_13

shut down, others have opted for online services, and other sectors have boosted their efforts to cover the country's needs during the crisis. Particularly, the greatest enhancements were seen in the technology sector since it is a sector in which innovation is rapid worldwide and it offers a wide arrange of services. These improvements are manifested in mobile applications, robotic machines, smart thermometers, boosting online services, and other digital tools that have increased in use with the pandemic spreading.

Furthermore, since the emergence of this virus coincided with the digital revolution that the world is experiencing in the twenty-first century, the amount of heterogeneous data generated as a result of this pandemic is following at an exponential rate. Moreover, this tremendous amount of data includes clinical data, patient information, news concerning the development of the virus around the world, the information's and statistics of COVID-19 cases by country … etc. [1]. Therefore, the processing of these data requires the use of artificial intelligence algorithms and Big Data analysis tools for several considerations, in particular visualization, analysis, and extraction of useful knowledge. Thus, the purpose of these applications is to develop intelligent systems of prevention and diagnosis and facilitate the process of finding drugs to treat COVID-19, and all that in an efficient and faster way compared to traditional treatment.

Machine learning is considered a type of artificial intelligence (AI) offering multiple algorithms and models to solve problems without specific computer programming. In the health-care context, the strength of artificial intelligence tools for processing Big Data is embodied in their ability to process various data and involve them in prediction for disease risk, diagnosis, prognosis, and appropriate treatments. Contrary to traditional bio-statistical methods, machine learning algorithms and deep learning models in health-care delivery are able to perform several tasks with high flexibility and scalability. However, such machine learning tools necessitate the accomplishment of certain tasks such as pre-processing, model training, and refinement of the system with respect to the actual clinical issue [2]. In this regard, researchers and health professionals have united their efforts, relying on artificial intelligence advances, analysis of Big Data by machine learning, and the latest medical developments to find the most effective solution to address the murderous disease. These efforts are reflected in vaccine and drug trials.

This paper discusses the past success of artificial intelligence and big data analytic in the treatment of diseases. It also presents the current advancements in machine learning tools as well as deep learning models for processing data sets related to the COVID-19 virus. In addition to that, the proposed work emphasizes the importance of Big Data models for treating diseases in the era of COVID-19. And finally, it gives a proposed analysis of the main factors that help decrease the development of the virus.

The remainder of the paper is organized as follows: Sect. 13.2 covers some relevant works presenting machine learning algorithms for treating infectious diseases in the past. Section 13.3 describes some healthcare applications obtained through advanced machine-learning tools in the COVID-19 era. Section 13.4 presents the requirements of the artificial intelligence model for efficient processing. Section 13.5 provides a proposed discussion and future directions. And finally, the conclusion is reported in Sect. 13.6.

13.2 Artificial Intelligence and Big Data for the Treatment of Infectious Diseases in the Past

In the 21st Century, Information and communications technology (ICT) has experienced remarkable development allowing access to massive and potentially relevant health care data available from a diversity of new sources such as national syndromic surveillance systems, genome databases, internet search key trends, vaccinology related data, Twitter data ... etc. [3]. Integrating big data into clinical and population health research necessitates in addition to new data sources, new thinking, advanced models for training, and new artificial intelligence tools [4]. Indeed, despite the availability of the health care data provided by different sources, these data are heterogeneous, unstructured, and contain inaccurate data that cannot be used directly for boosting patient healthcare. Hence, within the past few years, several improvements based on the artificial intelligence advancements have been suggested to extract the best clinical knowledge, reflected in multiple solutions aiming to address the complexity of the treatment of diseases by enhancing the efficiency of diagnosis, prevention, and drug discovery. Accordingly, the reviewed papers will be exposed chronologically.

Among the many uses of artificial intelligence in health care suggested at the beginning of the 21st Century, an application of syntactic pattern recognition which aims to diagnose disease symptoms occurring in the main pancreatic ducts and upper segments of ureters and renal pelvis as well as coronary arteries [5]. It performs this by analyzing morphological shapes of selected organs of the abdominal cavity and chest. Another intelligent healthcare application represented in a MERCURIO system intends to detect nosocomial infections, validate microbiological data and produces a real-time epidemiological information system [6]. In this direction, the suggested system includes three principal components. The first component describes the epidemiological database. The second component consists of applying the expert system ESMIS for real-time validation and monitoring. Hence, this knowledge-based system aims to analyze the microbiological information by several system modules and then avoid dangerous human and instrument-driven mistakes. In addition to that, data mining techniques have been applied with the purpose of automatically discover association rules from microbiological data, as well as obtain alarm rules from them. The last component represents a statistical module used for performing statistical analyses and identifying outbreaks. The tests conducted on the MERCURIO system considering real microbiological data demonstrate its efficiency for knowledge discovery and its reliability since it achieves an accuracy of 98.5%.

In the same context, an additional suggested solution represented in a mapping process having as objective to limit the clinical and epidemiological problems by predicting spatial disease risk for dengue disease. Thus, the mapping process begins by gathering the disease occurrence records, and then it produces putative pseudo-absence points based on the extent of infectious disease occurrence and the assemblies of known occurrence. The next step consists of collecting a range of epidemiologically relevant environmental covariates. Thereafter, the process involves the accelerated regression tree technique, which is statistically flexible and ensures beneficial comparative reviews of efficiency. The last step intends to predict the probability of disease occurring at each location for the sake of generating a risk map with a quantified measure of uncertainty. The big data approach is implicated in every step of the disease mapping process such that it can be more enhancing and evolving risk maps [7].

From a different perspective, Big Data analytic has a significant role in personalized healthcare and biomedical discovery. Diverse significant studies were conducted to emphasize the importance of this approach such as [4, 8, 9]. The Collaborative Assessment and Recommendation Engine (CARE) [4] is one of the proposed data mining techniques aiming to raise disease modeling and understanding by producing a personalized disease risk profile. In this regard, the suggested framework starts by comparing the medical history of an individual with the other patients' medical histories. The next step attempt to select the most similar patients to the patient under consideration based on defined similarity constraints. Then, collaborative filtering using inverse frequency and vector similarity is applied to the selected patients to obtain a ranked list of diseases ordered from highest to lowest risk. The purpose is then to generate a generalized predictive model about possible disease risks in the future. In a similar vein, diverse realizations of Artificial Intelligence techniques in the field of cardiology were conducted to emphasize the importance of a personalized healthcare approach. In this direction, artificial neural networks are employed in cardiac imaging to provide accurate interpretation based on patient electronic medical records for diagnosis and treatment of different cardiac diseases. These networks take advantage of various approaches such as gradient-based training, fuzzy logic, Bayesian methods, and genetic algorithms to ensure cost-effective personalized medicine [9]. Furthermore, the implementation and the application of personalized medicine require powerful big data solutions combining software and hardware tools for deep analysis and best exploitation of 'omics' and clinical health data [10].

13.3 Artificial Intelligence and Big Data for Fighting COVID-19

Artificial intelligence has become essential in almost all fields, and particularly in health. In this regard, it requires big clinical data for prevention, diagnosis, or outright finding suitable treatment or vaccines for infectious diseases. In the face of the global threat caused by the SARS Coronavirus and MERS Coronavirus, several research and studies were carried out, based on artificial intelligence techniques with regard to proposing adequate solutions required for counteracting the global crisis accurately and speedily.

13.3.1 Prevention Techniques

Accordingly, among the solutions proposed, a preventive system having as a goal to detect corona virus from earlier clinical symptoms [11]. The system suggested aims to classify textual clinical reports into four disease categories using various supervised machine learning algorithms as well as ensemble learning techniques. Thus, the process starts by collecting patients' data sets, then extract clinical notes and findings. The next phase focuses on five essential steps aiming to refine the unstructured data in order to obtain satisfactory accuracy after classification. Following the pre-processing steps, the feature engineering stage consists of selecting relevant features and gives their corresponding weights. The last stage performs classification with traditional machine learning algorithms and ensemble learning models. The machine learning algorithms include logistic regression and Multinomial Naive Bayes algorithm which achieve better accuracy than the other algorithms. Therefore, the framework above-exposed can be useful for examining the clinical reports and take convenient decisions.

In the same context, the authors in [12] introduce a framework for predicting mild patients with malignant progression using both the clinical indicators and CT data for COVID-19 transformation prognosis. Thus, the process starts by collecting the input data to be processed that is composed of static and dynamic data. The static data contain clinical data and information of patients while the dynamic data include quantitative chest CT data accumulated at different times. The dynamic data are thereafter flattened to form the appropriate CT feature vector. Simultaneously, the system applies a multi-layer perceptron (MLP) to the static data in order to obtain a feature vector with the same dimension produced from the dynamic data. The next step consists of merging the static and dynamic data to get the input of long short term memory (LSTM). The output of LSTM is then fed into completely connected network layers. The last step aims to display the probability of patient alteration to the severe/critical stage by adding a Softmax layer at the top of the network. The suggested predictive model is useful to improve the treatment process and consequently reduce mortality by the early detection of infectious diseases.

Another framework developed by [13] aims to predict patients with acute respiratory distress syndrome (ARDS) based on machine learning algorithms, such as decision trees, random forests, and support vector machines. In that respect, the initial step involves the collection of historical data from patients having a positive result for the Coronavirus test. The next step consists of selecting the most predictive features for ARDS by applying feature selection methods that include filter methods and wrapper methods. The filter methods employed are based on entropy. These methods are manifested in the Information gain method, Gini index method, and Chi-Squared statistics. Whereas, the wrapper method is based on an iterative process starting with an empty set of features and then adding progressively the features as long as the framework performance is enhanced. Once the quality of the framework is no longer increasing, the process stops. The last step requires the execution of predictive models with the best predictive features in order to identify the patients who will develop acute respiratory distress syndrome (ARDS). Better accuracy is achieved with decision trees, random forests, and support vector machines machine learning algorithms that learn from previous experiences for more precise and successful prediction.

13.3.2 Diagnostic Techniques

As is the case for prevention techniques, an early and firm diagnosis for covid-19 based on machine learning and deep learning algorithms can be efficient and decisive to improve the treatment of the diseases and then reduce mortality risk. Several architectures were designed to detect the covid-19 virus from various medical images.

Among these architectures, the model proposed in [14] for the detection of COVID-19 by analyzing medical images based on Generative Adversarial Networks (GAN), and deep learning models. In that regard, it consists of three fundamental stages. The first stage aims to generate new images from the input datasets based on three strategies including the generator strategy, the augmentation strategy, and the discriminator strategy. These strategies compose the primary steps of GAN architecture. The second stage aims to split the images produced into two reasonable parts. The first part comprises 90% of the original dataset and is dedicated to the training and validation phase, while the second part comprises only 10% of the original dataset and is dedicated to the testing phase. Thereafter, the process aims to apply Resnet50, Shufflenet, and Mobilenet deep learning models for validation and training. The last stage classifies the obtained datasets into the Covid-19 class and the normal class. As a final result of the performance evaluation, the Shufflenet model is chosen to be the most efficient in terms of time, memory consumption, and CPU usage.

In a similar vein, the authors in [15] present a deep machine learning approach for detecting COVID-19 accurately in India using Deep Neural Network models on Chest Radio-graph datasets. Thus, the system requires the collection of COVID-19 Chest X-ray datasets as a first step and then applies a pre-processing procedure to the gathered datasets to eliminate outlier images. The next step involves the use of a heatmap filter by performing easy feature extraction and scaling, in order to improve the model quality. Thereafter, the process consists of partitioning the processed images into two separate parts. The first part includes the training images while the second part composes the testing images. In this regard, the training images are introduced subsequently into different deep learning models for faster training and to efficiently classify the features of the radio-graphs. The implemented Convolution

Neural Network (CNN) based algorithms include Inception V4, VGG 19, ResNet V2 152, and DenseNet. Finally, after evaluating the performance of the implemented algorithms, the DenseNet is selected to be the most efficient model in terms of accuracy, and therefore it is a suitable classifier in real scenarios.

Another contribution proposed by [16] intent to predict the risk of positive COVID-19 diagnosis using diverse machine learning algorithms. In this direction, the process starts by collecting clinical data from patients who performed the RT-PCR tests. Secondly, the process consists of dividing a randomly selected sample into two parts. The first part contains 70% of the full sample and is dedicated to training the machine learning algorithms. While the second part comprises 30% of the full sample and is devoted to evaluating the quality of the models on new datasets. The machine learning algorithms applied include neural networks, random forests, gradient boosting trees, logistic regression, and support vector machines. The performance evaluation based on tenfold cross-validation demonstrates that the support vector machines algorithm is the best predictive algorithm.

13.3.3 Drugs Discovery and Repositioning

Drug repositioning is making an important contribution to the healthcare domain. Indeed, it consists of the reuse of old drugs to develop suitable ones for new infectious diseases, orphan diseases, and other serious diseases for which the treatment is not available. Contrary to the conventional drug development approach which requires 10–15 years as well as a high cost of financing for discovering new drugs, the drugs repositioning approach is cost-effective in terms of time, cost, and efforts [17].

In that respect, several studies and clinical trials were conducted to test some popular drugs and consider them as candidates for treating the virus caused by SARS-CoV-2, including [18]. These various drugs have shown their effectiveness in reducing the viral load of SARS-Cov-2 each one of them separately. Furthermore, the use of artificial intelligence, powerful calculation algorithms, and prime machine learning (ML) theories has become crucial to make the drug repositioning approach more effective. To this end, a Drug Database and Open Chemical/Drug Database are processed with artificial intelligence strategies to get desired drugs.

Distinct companies have adopted such techniques for the approach above mentioned such as BenevolentAI company, Innoplexus company, Deargen company, and VantAI, just to name a few. The use of artificial intelligence systems enabled these organizations to anticipate the anti-parasitic particles able to block the infection movement and understanding the behavior of viral proteins toward human proteins. Hence, the effective exploitation of artificial intelligence techniques leads to identifying the most balanced approach to managing the disease. Also, artificial intelligence is employed with computational Drug designing for purposes of improving access to treatment for new diseases. Absolutely, computational Drug designing is a strong technique in medicinal chemistry. This approach aims to analyze the molecular structure, properties, interactions, and behaviors of drugs with specialized functions.

Thus, in particular, the examination of molecular properties allowed recognizing the starting points as successful molecules for Covid-19 [17].

13.4 Requirements of Artificial Intelligence Model for Processing Big Clinical Data

Processing big clinical data using machine learning algorithms and deep neural networks depends on a thorough process. Such a process is composed of several steps starting by collecting the appropriate data and finishing by making the necessary clinical decisions. In particular, in the healthcare domain, this approach depicts the steps illustrated in Fig. 13.1. Thus, the first step of the exposed model involves gathering various clinical data comprising real-world clinical data, imaging data, laboratory findings, and demographic data. This step requires the availability of the clinical data sets which can be gathered with different capacities, in particular directly from patients by healthcare professionals, from intelligent devices or mobile applications. This in turn demands powerful hardware and software tools to guarantee better storage capacity. The first step is followed by a pre-processing step

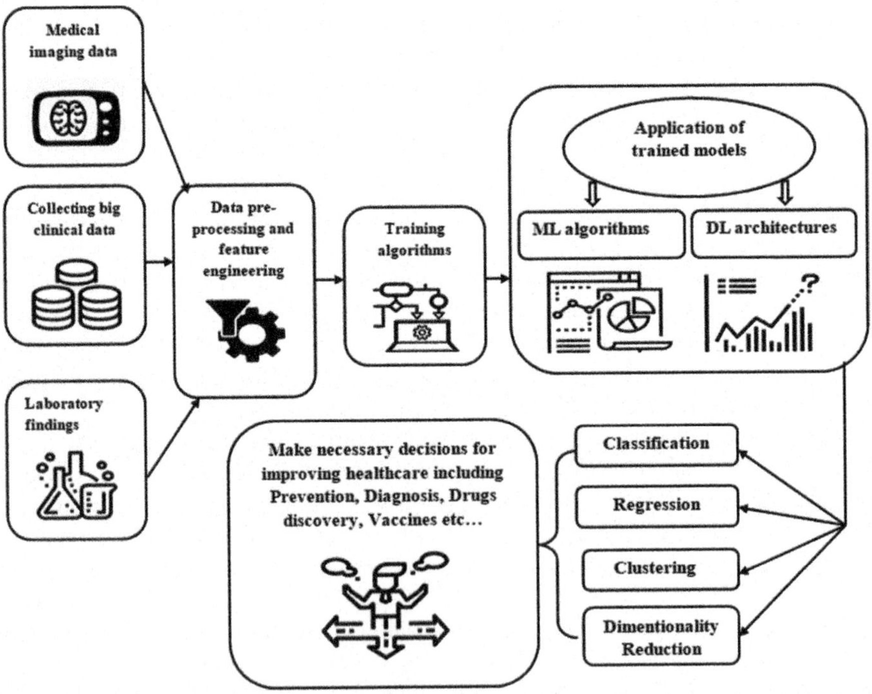

Fig. 13.1 The main steps for processing big clinical data with machine learning tools

which includes multiple operations for example data cleaning, data transformations, and data reduction. Thereafter, feature engineering techniques are applied to more improve the quality of the pre-processing strategies. Next, and if necessary split the processed data into two parts, one part is consecrated for training on data sets, whereas the second part is dedicated to testing data sets. The next step aims to train the machine learning algorithms with the available clinical data sets which allow these algorithms to learn from previous experiences and therefore increase their learning rate. Amongst the various factors which affect the learning performance of the deep learning models, the amount of data used, the selection of the type of variables, and the number of variables. The machine learning algorithms and deep learning architectures are then applied according to their outcomes embodied in regression, classification, clustering, and dimensionality reduction. Indeed, the most useful algorithms for the purposes of the clinical decision include, but are not limited to, linear regression, logistic regression, decision tree, support vector machines, random forests, and gradient boosting trees. Similarly, the deep learning architectures manifested in Resnet50, Shufflenet, Mobilenet, CNN, and RNN, etc. In general, the objective of this process is to enhance the quality of diagnosis, prevention, finding suitable drugs or vaccines for infectious diseases, and thus make the necessary clinical decisions.

13.5 Discussion and Future Directions

With the onset of a pandemic, several factors can help make big data solutions and artificial intelligence tools more efficient in terms of time and performance in order to combat the pandemic encountered. Among these factors, there is the availability of clinical data, the quantity, and quality of the data collected, the ability to learn the algorithms adopted for data processing, etc. Indeed, the reliability and availability of the data collected play a primary role for the intelligent model chosen, since the clinical decision made after processing these data will be extremely based on clinical data retrieved from different sources. In this context, the efforts of researchers, experts, and health professionals must be united to ensure the quality and reliability of the large clinical data collected. Likewise, the machine learning algorithms must undergo high learning by increasing the percentage of clinical data dedicated to the training of these algorithms and consequently to obtain an efficient and rapid clinical decision.

From another perspective, diseases have risk indicators that can be analyzed and exploited in personalized medicine. In this sense, studying the behavior of the diseases encountered and comparing them to the clinical data relating to each patient can help to improve the ability to prevent the adverse effects of the disease to which such a patient is exposed. This calls for real intelligent systems able to carry out these tasks in a reduced time for a better performance of prevention and diagnosis. In addition to this with the spread of a deadly disease caused by a virus, it is essential to have an intelligent system that mimics human behavior using deep learning algorithms to study the comportment of virus molecules towards the human body and therefore accelerate the process of reuse and discovery of drugs as well as research of suitable vaccines.

The spread of the deadly pandemic caused by the SARS-Cov2 virus is an example of a global crisis for which the tools of artificial intelligence and big data analytics are decisive solutions to improve the performance of clinical decision-making. But despite the significant development that this field has experienced, it still requires continuous development until it reaches the point of not losing anyone's life due to such a virus. Among the solutions proposed is the fact of being able to prevent the arrival of a pandemic by the artificial intelligence systems, which makes it possible to guarantee a rapid reaction from researchers and health professionals to stop the spread of the pandemic encountered. Another suggestion consists in the possibility of developing a potent vaccine capable of combating any infectious virus by taking advantage of the ability of intelligent systems to mimic the reaction of the human body to the attack it suffered by studying all of the involved factors.

13.6 Conclusion

The exploitation of artificial intelligence frameworks for processing big data related to new infectious diseases can improve the performance of the prevention, diagnosis, and even discovering suitable drugs and vaccines in the future. This paper presents the achievements of the machine learning algorithms and deep learning networks for the treatment of infectious diseases within the past years and in the COVID-19 era. Moreover, the suggested work emphasizes the requirements of artificial intelligence models to make suitable clinical decisions. In the end, it gives a proposed discussion of the main factors that can help decrease the development of an encountered pandemic in the future. In our future research, we intend to analyze the intensity of the harmful effects of the virus caused by SARS-CoV-2 using machine learning algorithms and based on different indicators and therefore take advantage of this analysis to stop the spread of the virus. Furthermore, the objective would be to use the results of the analysis carried out to fight against pandemics that could conquer the world in the future.

References

1. Jiang, X., et al.: Towards an artificial intelligence framework for data-driven prediction of coronavirus clinical severity. Comput. Mater. Contin. **62**, 537–551 (2020). https://doi.org/10.32604/cmc.2020.010691
2. Chawla, N.V., Davis, D.A.: Bringing big data to personalized healthcare: a patient-centered framework. J. Gen. Intern. Med. **28**(3), 660–665 (2013). https://doi.org/10.1007/s11606-013-2455-8

3. de M. Batista, F., Miraglia, J.L., Donato, T.H.R., Filho, A.D.P.C.: COVID-19 diagnosis prediction in emergency care patients: a machine learning approach. medRxiv, p. 2020.04.04.20052092 (2020). https://doi.org/10.1101/2020.04.04.20052092
4. Bai, X., et al.: Predicting COVID-19 malignant progression with AI techniques. Social Science Research Network, Rochester, NY, SSRN Scholarly Paper ID 3557984 (2020). https://doi.org/10.2139/ssrn.3557984
5. Costa, F.F.: Big data in biomedicine. Drug Discov. Today **19**(4), 433–440 (2014). https://doi.org/10.1016/j.drudis.2013.10.012
6. Ogiela, M.R., Tadeusiewicz, R.: Artificial intelligence structural imaging techniques in visual pattern analysis and medical data understanding. Pattern Recogn. **36**(10), 2441–2452 (2003). https://doi.org/10.1016/S0031-3203(03)00089-X
7. Hay, S.I., George, D.B., Moyes, C.L., Brownstein, J.S.: Big data opportunities for global infectious disease surveillance. PLoS Med. **10**(4), e1001413 (2013)
8. Detection of COVID-19 Using Chest Radiographs with Intelligent Deployment Architecture | SpringerLink. https://doi.org/10.1007/978-3-030-55258-9_7. Accessed 10 Apr 2021
9. Khanday, M.U.D., Rabani, S.T., Khan, Q.R., Rouf, N., Mohi Ud Din, M.: Machine learning based approaches for detecting COVID-19 using clinical text data. Int. J. Inf. Technol. **12**(3), 731–739 (2020). https://doi.org/10.1007/s41870-020-00495-9
10. Lamma, E., Mello, P., Nanetti, A., Riguzzi, F., Storari, S., Valastro, G.: Artificial intelligence techniques for monitoring dangerous infections. IEEE Trans. Inf. Technol. Biomed. **10**(1), 143–155 (2006). https://doi.org/10.1109/TITB.2005.855537
11. Krumholz, H.M.: Big data and new knowledge in medicine: the thinking, training, and tools needed for a learning health system. Health Aff. (Millwood) **33**(7), 1163–1170 (2014). https://doi.org/10.1377/hlthaff.2014.0053
12. Ngiam, K.Y., Khor, I.W.: Big data and machine learning algorithms for health-care delivery. Lancet Oncol. **20**(5), e262–e273 (2019). https://doi.org/10.1016/S1470-2045(19)30149-4
13. Wong, Z.S.Y., Zhou, J., Zhang, Q.: Artificial Intelligence for infectious disease big data analytics. Infect. Dis. Health **24**(1), 44–48 (2019). https://doi.org/10.1016/j.idh.2018.10.002
14. The Detection of COVID-19 in CT Medical Images: A Deep Learning Approach | SpringerLink. https://doi.org/10.1007/978-3-030-55258-9_5. Accessed 10 Apr 2021
15. Dilsizian, S.E., Siegel, E.L.: Artificial intelligence in medicine and cardiac imaging: harnessing big data and advanced computing to provide personalized medical diagnosis and treatment. Curr. Cardiol. Rep. **16**(1), 1–8 (2014). https://doi.org/10.1007/s11886-013-0441-8
16. Mohanty, S., Harun AI Rashid, M., Mridul, M., Mohanty, C., Swayamsiddha, S.: Application of artificial intelligence in COVID-19 drug repurposing. Diabetes Metab. Syndr. Clin. Res. Rev. **14**(5), 1027–1031 (2020). https://doi.org/10.1016/j.dsx.2020.06.068
17. Coronavirus disease (COVID-19)—World Health Organization. https://www.who.int/emergencies/diseases/novel-coronavirus-2019. Accessed 10 Apr 2021
18. Low, Z.Y., Farouk, I.A., Lal, S.K.: Drug repositioning: new approaches and future prospects for life-debilitating diseases and the COVID-19 pandemic outbreak. Viruses **12**(9), Art. no. 9 (2020). https://doi.org/10.3390/v12091058

The manufacturer's authorised representative in the EU is Springer Nature Customer Service Centre GmbH, Europaplatz 3, 69115 Heidelberg, Germany. If you have any concerns regarding our products, please contact ProductSafety@springernature.com

Printed and bound by CPI Group (UK) Ltd, Croydon, CR0 4YY
30/09/2026
02225310-0003